AF575668

A Just Transition

Making Energy Poverty History with an Energy Mix

NJ Ayuk

Made for Success Publishing
P.O. Box 1775 Issaquah, WA 98027
www.MadeForSuccess.com

Distributed by Made for Success Publishing

First Printing

Library of Congress Cataloging-in-Publication data
Ayuk, NJ
A Just Transition: Making Energy Poverty History with an Energy Mix
p. cm.

ISBN: 978-1-64146-750-6 *(HDBK)*
ISBN: 978-1-64146-751-3 *(eBOOK)*
ISBN: 978-1-64146-752-0 *(AUDIO)*

Printed in the United States of America

For further information contact Made for Success Publishing
+14255266480 or email service@madeforsuccess.net

Contents

Preface

REMEMBERING A FRIEND, A MENTOR, AND ONE OF THE AFRICAN ENERGY INDUSTRY'S GREATEST ALLIES

Mohammad Sanusi Barkindo's appointment as Secretary General of the Organization of the Petroleum Exporting Countries (OPEC) in 2016 was incredibly good news.

OPEC's new leader brought extensive oil industry knowledge to the table and was already admired for his diplomacy, his brilliant mind, and his devotion to public service. And he was an African. For decades, Saudi Arabia—with the most oil reserves among OPEC's 13 member nations[1]—had dominated the organization's production decisions. But in his role as Secretary General, Barkindo adjusted the organization's focus. He never discounted the views and priorities of the seven African member countries in OPEC, understanding the power of strategic oil and gas production to help eradicate African energy poverty and build solid, diverse African economies.

Up until his sudden death at age 63 on July 5, 2022—only 3 weeks before his planned retirement from OPEC—Barkindo remained an unwavering advocate for just energy transitions for developing nations, including those in Africa. His voice and perspectives were greatly needed in our continent and around the globe, and his loss will be felt for years to come.

His death creates a painful void for me as well. It is difficult to fully describe my deep affection and respect for Secretary General Barkindo. He was a mentor and a dear friend, one of the few people in my life I spoke with multiple times daily. It meant the world to me when he wrote the foreword to my last book. I'm heartbroken that this book opens with a tribute in his memory.

As I said shortly after receiving the tragic news of my friend's death in Africa, Secretary General Barkindo will always be remembered as a patriot, an instrumental figure in the fight for the continent's right to develop our oil and gas. He will be honored as the man who united producers, helped create OPEC+, fought to alleviate energy poverty, and strengthened Africa's position as a global energy supplier.

An African Success Story

Mohammad Sanusi Barkindo was born in 1959 in Northern Nigeria's impoverished Adamawa State. He was a child during the civil war that took place there between 1967 and 1970, an experience that some say contributed to his commitment to diplomacy.[2]

He prepared well for a future in politics and oil industry leadership. Barkindo had a bachelor's degree (with honors) in political science from Ahmadu Bello University in Zaria, Nigeria; a postgraduate diploma in petroleum economics from Oxford University; and an MBA in finance and banking from Washington University. He also was a Fellow at George Mason University in Fairfax, Virginia (2013-2016) and received an honorary doctorate in science from the Modibbo Adama University, Yola, a research university in Northeastern Nigeria.

During the early years of his career, Barkindo worked for the Nigerian Mining Corporation and as a special assistant to Nigeria's Minister of Mines, Power & Steel at the time, Rilwanu Lukman.[3] Barkindo went on to hold various posts within the Nigerian government, including the state-owned Nigerian National Petroleum Corp. (NNPC), which he eventually led as its group managing director/CEO from January 2009 to April 2010. He also served as Nigeria's governor for OPEC, Nigeria's representative on OPEC's Economic Commission Board, and OPEC's acting secretary general.[4]

A Legacy of Building Bridges

During his years at the helm of OPEC, Barkindo earned a reputation for helping leaders with diverse goals to find common ground.

American diplomat Paula Jon Dobriansky, who worked closely with Barkindo when she served as Under Secretary of State for Global Affairs, put it this way: "He was a thoughtful leader, a strong consensus builder, and a tenacious problem solver."[5]

Barkindo demonstrated all of those qualities in 2016 when he played a major role in creating OPEC's landmark Declaration of Cooperation (DoC). Under that agreement, 24 oil-producing countries (both OPEC members and non-members) came together to rescue the global oil market. In response to surges in shale oil production, DoC participants agreed to cut their production to stabilize the oil market.

That alliance has remained intact in the form of OPEC+, which agreed to oil-industry-saving production cuts again in 2020 after COVID-19-related lockdowns slashed oil and gas demand.

On the day of Barkindo's death, after he'd arrived in the Nigerian capital of Abuja to speak at an energy summit, Nigerian President Muhammadu Buhari recognized Barkindo's contributions to establishing the DoC. "Your time in charge of the affairs of OPEC has been a very challenging one for the global oil industry," Buhari told

Barkindo. "Oil producers were finding it difficult to come together to address challenges that were crippling the oil market. There is no doubt about your efforts in putting together the Declaration of Cooperation, which is the largest cooperative effort in the history of OPEC and the global oil industry and also the longest in duration in the history of the organization. This was a Herculean task."

"A Much-Loved Leader"

It's fair to say that Barkindo was admired not only for what he accomplished but also for how he did it. He was admired for his passion for the oil industry and for the way he treated others. His character is reflected in the reluctant goodbyes and tributes issued by global leaders and energy industry representatives since Barkindo's death.

"He was the much-loved leader of the OPEC Secretariat, and his passing is a profound loss to the entire OPEC family, the oil industry and the international community," said the Secretariat in a statement shortly after Barkindo's death.

> *Throughout HE [His Excellency] Barkindo's long career, there have been several central themes that have driven him: an infectious passion for the petroleum industry; an unwavering belief in oil's poverty-eradicating potential; a steadfast commitment to sustainable development; the importance of dialogue and multilateralism; and, most fundamentally of all, treating everyone with respect and kindness.*

Yesar Al-Melaki, an analyst at MEES Energy, tweeted about the strong impression Barkindo made on him. "First time I met Mohammad Barkindo I was a nobody, just a fresh energy economics graduate," Al-Melaki wrote. "Talking to him, he put me immediately at ease, asked for my business card—I didn't have any. He offered his, gave me another one and asked me to write my info on the back."[6]

African Energy Industry Ally

I, too, respected Barkindo for his wisdom, leadership, and the way he treated those he interacted with. But I cannot write about his legacy without emphasizing how much he meant to those of us in Africa. Because of his support for the continent's energy industry, the Africa Oil & Power Conference named Barkindo their Africa Oil Man of the Year in 2018.[7]

His willingness to champion Africa's energy industry would become even more important in the years that followed. In this book, you'll read about the unrelenting pressure African oil- and gas-producing nations have been facing in recent years to leave their petroleum resources in the ground in the name of protecting the climate. Secretary General Barkindo stood by our continent's energy industry when we needed him most. He didn't hesitate to urge African nations to continue producing oil. OPEC, he said, would "assist in whatever way possible."

This was particularly notable during an August 2021 meeting at the Republic of the Congo's National Oil Company, Société nationale des pétroles du Congo (SNPC), part of an OPEC and African Energy Chamber delegation to the country. Barkindo made it clear that while renewable energy sources like wind and solar power have valuable roles to play in Africa's future, they won't be enough to meet Africa's vast and growing energy needs.

"OPEC categorically rejects the narrative that the energy transition is from hydrocarbons to renewables," Barkindo said. "All sources of energy are required today in order to meet the challenges of climate change and the future demand for energy."[8]

Secretary General Barkindo got it. We in the African energy industry will miss our friend and advocate.

Nigerian Minister of State for Petroleum Resources Timipre Sylva wrote about what Barkindo meant to him and Nigeria in a letter to the late OPEC leader's family. Secretary General Barkindo was a "worthy ambassador of our great country Nigeria," Sylva wrote. "His contribution to the global energy space will be difficult to match."[9]

Just a week earlier, Sylva wrote to Barkindo himself to congratulate him for the successful completion of his service to OPEC, from steering the DoC forward to managing the effects of the COVID-19 pandemic. "Your tremendous and unparalleled contribution to the OPEC during the past six years has been mind-boggling," Sylva wrote to Barkindo. "It is globally acknowledged that you have done much for the stability of the oil market during your tenure... My dear brother, your gentle and unperturbed demeanor is impressive and worth emulation. Like the proverbial saying, You came, you saw, and you conquered. You have earned yourself a place in the hall of fame for giants in the petroleum industry."[10]

Gabriel Mbaga Obiang Lima, Equatorial Guinea's Minister of Mines and Hydrocarbons, said he will always be grateful for Barkindo's mentorship and his example as a leader. "Being with him was like earning an MBA," Lima said, adding that OPEC's African member countries, including Equatorial Guinea, owe Barkindo a debt of gratitude as well for amplifying their voices.

"He made Africa relevant in OPEC," Lima said.

Bruno Jean-Richard Itoua, Congo's Minister of Hydrocarbons, said he will remember Barkindo for his in-depth understanding of the oil and gas industry and his respect for those who contribute to its success. "I'll never forget the time I spent with His Excellency Mohammad Sanusi Barkindo when he toured Congo: the first-ever visit to our country by an OPEC secretary general," Itoua said. "He was extremely interested in our petroleum industry's potential and progress up to that point. He also took the time to meet with the Prime Minister, the President of the Senate, the President of the House of Representatives, the Secretary General of the African Petroleum Producers Organization, and representatives of some of the international oil companies working with our national oil company (Société nationale des pétroles du Congo)."

That was not at all unusual for Secretary General Barkindo. He was about building bridges and capitalizing upon opportunities; he never saw the private sector as the enemy. As a result, he succeeded not only in giving OPEC's African member nations a voice but also in bringing Africa's private sector to the table. He never failed to consider their unique perspectives and ideas.

"To say Secretary General Barkindo was a hard worker would be an understatement," said Angolan Minister of Mineral Resources and Petroleum, Diamantino Pedro Azevedo. "But it's important to understand that his underlying motivation wasn't to get things done; he saw the good left to be achieved, particularly, the positive impact that thriving oil and gas industries could have on developing countries, and he was determined to help bring that vision to fruition."

His drive and optimism were contagious. Secretary General Barkindo was the kind of leader who inspired—and welcomed—others to work alongside him, even former U.S. President Donald Trump, once an outspoken critic of OPEC. Barkindo and Trump worked cooperatively to forge the 2020 OPEC+ deal that I spoke of earlier, an agreement that helped steer the oil industry away from disaster in the face of the COVID-19 pandemic.

In January 2022, I wrote about Barkindo's legacy in response to his plans of turning the reins of OPEC over to his successor, Haitham Al-Ghais, in July.

Barkindo, throughout his tenure with OPEC, has fought the good fight, finished his race, and kept the faith... We knew the day was coming... Yet the industry has become accustomed to Mr. Barkindo's steady hand on the rudder, guiding OPEC through the volatile waters that global oil and gas producers must navigate, including growing public sentiment against fossil fuels.

If he had just maintained order during his term, that would have been enough, especially given the upheaval brought on by the pandemic. But

> *Barkindo (has never been) one to be satisfied with the status quo, even during the most trying times.*[11]

And while it's painful to re-read this, now that our goodbye is about the man's lifelong legacy and not just his retirement, the words are truer than ever.

Chapter 1

AFRICA NEEDS ENERGY—AND MORE THAN ONE KIND OF ENERGY

AFRICA IS FACING a serious problem: energy poverty.

As I write this book, the continent of my birth is still home to millions of people who have no access to electricity: 600 million, according to the International Energy Agency (IEA).[1] This lack of access isn't a trivial matter. Without electricity, these 600 million Africans are simply not in a position to achieve the same levels of health and wealth that prevail in more developed countries. Moreover, without electricity, they will struggle individually and collectively to take the steps needed to achieve economic development.

But solving the problem of energy poverty is not simple.

We now know that crude oil and natural gas—which offer a better balance of energy density, economy, and ease of use than any other known fuel sources—cause pollution and generate emissions that contribute to climate change.[2] Thanks to new refining techniques and tighter quality standards, humanity has made headway with respect to reducing pollution.

However, climate change is proving to be a more difficult conundrum. World leaders are still trying to figure out how (and whether) we can reduce the carbon dioxide emissions that are the inevitable product of extracting and burning fossil fuels without giving up the economic benefits that fossil fuels and their derivatives offer. World leaders have worked out a plan, in the form of the Paris Agreement,[3] which is designed to keep average global temperatures from rising by more than 1.5 degrees Celsius in the long term.

It is far from clear that this plan will work.

If this plan doesn't work, Africa stands to bear a disproportionate share of the consequences of climate change.[4] Since the continent is at a lower level of economic development than its developed neighbors, it exhibits less resilience in the face of climate-related adverse events such as rising sea levels, storms, flooding, and drought, all of which may lead to food insecurity or large-scale population movements. And since a larger share of Africa's population is involved in farming, a larger share of its population runs the risk of being displaced or put out of work if land becomes unproductive.

In short, we Africans have every reason to seek to minimize climate change. It behooves us to understand that if the world doesn't contain carbon dioxide emissions to curb the rise of global temperatures, Africa will suffer greatly.

But Africa is already suffering: As I noted above, the IEA puts the number of people without electricity in Sub-Saharan Africa at 600 million. But that's not the whole story. 600 million people amounts to three-quarters of the global population who lack reliable power, approximately 940 million people.[5] [6]

Africa bears a disproportionate share of not only the anticipated future problems of climate change, but the present-day, here-and-now problem of energy poverty.

So what should we do?

Should we prioritize the goal of preventing climate change over the goal of addressing energy poverty? Or should we focus on ways to deliver power to the hundreds of millions of Africans who currently lack access to electricity?

Obviously, we should try to achieve both objectives. Both are important. But how do we do this? And which should come first?

This book is my attempt to provide an answer to those questions.

Should Africa Put Renewable Energy First?

The answer put forward by the IEA (and others) is that renewables are the answer to Africa's energy poverty—that solar panels, wind turbines and the like can supply the continent with the electricity it needs without increasing its contribution to emissions, which is currently quite small.[7]

My response is this: It would be nice if that were true. I would be thrilled if Africa was in a position to wipe out energy poverty with nothing but renewable energy. But it's not. As of this writing, the continent cannot make the transition away from fossil fuels independently or immediately.

This is because solar, wind, and other types of renewable and alternative energy projects cost more than most African countries can bear to pay, and require more specialized technologies than most African countries can provide.

The global community has offered to help the continent make the transition to renewable energy. More specifically, in 2009, the world's more developed countries came together and committed to making USD100 billion per year in climate financing available to less developed countries by 2020.[8] To date, though, the states that made that pledge have consistently fallen short, and as a result, Africa and other developing regions are still waiting to receive assistance at the level they were promised.[9]

In all honesty, during this period of waiting, I have become increasingly uncomfortable with the idea that Africa's approach to the energy transition ought to be linked so firmly to the actions of outside powers. That is, I've begun to wonder whether there isn't a whiff of colonialism[10] about the prospect of African countries remaining on hold, pining for solar panels and wind turbines until such time as they are allowed access to solutions handed down from on high.

But it doesn't have to be like this.

Africa can approach the energy transition in a different way—its own way.

Africa can move forward in a manner that allows it to develop its own energy capacity to the greatest extent possible, on every front. This means making the most of *every* energy resource available—not just sun and wind, but also fossil fuels, especially natural gas. Let's make "both-and" instead of "either-or" our slogan!

In practical terms, that means focusing most of our efforts on natural gas at the start of this transition process and phasing renewables in over time.

Why Focus on Gas First?

So why should Africa make gas development its first major priority on the path toward its energy transition?

I can think of at least one compelling reason: the sheer quantity of Africa's reserves of natural and associated gas are substantial. They may not be quite as vast as those of Russia or the Middle East, but they are certainly very large, measuring 22.7 trillion cubic meters (tcm) altogether (802 trillion cubic feet),[11] with about three-quarters of the total located in Sub-Saharan Africa. Moreover, these reserves are probably even larger than we currently realize, as some unconventional and offshore basins remain unexplored or under-explored.[12]

Nigeria's experience on this front is intriguing. In mid-2021, the West African country's Minister of State for Petroleum Resources,

Timipre Sylva, noted that Nigeria had "accidentally" discovered so much gas while looking for oil that it had been able to raise its proven gas reserves to more than 5.7 tcm altogether (200 trillion cubic feet). He also raised the possibility that additional exploration might be similarly fruitful, bringing Nigeria's reserves all the way up to 17 tcm (600 trillion cubic feet).[13]

Even in the absence of such speculation, however, Africa is known to possess large amounts of gas—and what's more, its proven resources are not confined to a single geographic area.

Rather, these resources are distributed over multiple parts of the continent, as nearly half of Africa's 55 recognized countries are known to possess gas.[14] I believe this is important, as it raises the possibility that African countries can develop resources not just for their own use, but also for delivery to nearby markets, rather than mostly or exclusively for export.

In short, Africa has quite a lot of this type of fuel and, as such, it ought to give serious consideration to using more—much more—of these sources to generate electricity. But gas isn't just attractive because of its abundance. It's also attractive because it has high energy density and makes use of well-known, low-cost, high-efficiency technologies that African countries can easily access. Additionally, it also has the potential to create jobs—both directly, in the oil, gas, and petrochemical industry, and indirectly, in the industrial, manufacturing, and related sectors.

Crucially, given that Africa cannot afford to ignore climate change any more than it can afford to ignore energy poverty, gas is also attractive because it has a lower environmental impact than other fossil fuels. I'm not denying that gas generates carbon dioxide and methane emissions; however, we shouldn't lose sight of the fact that its carbon footprint is smaller than that of coal or petroleum products,[15] both of which are currently being used in many of Africa's existing thermal power plants (TPPs). Additionally, it's worth remembering that there are technologies that can reduce emissions, and that there are carbon credits that can offset emissions.

I'd also like to point out that gas has a solid track record of playing a major role in reducing emissions in developed economies. According to the U.S. Energy Information Administration (EIA), switching TPPs from coal to gas was the main reason why the U.S. power-generating sector saw carbon dioxide emissions sink by 32% between 2005 and 2019,[16] even as the U.S. economy grew by 28%.[17] Consequently, African gas produced for export will be able to support efforts to reduce global emissions, even as the gas that stays on the local market helps minimize Africa's own carbon footprint as more power plants come online.

In practical terms, if we want to end energy poverty in Africa—if we want to put an end to the days when hundreds of millions of Africans depend on inefficient biofuels such as wood and charcoal, which contribute to deforestation and pollution[18]—the best way to do so quickly will be to bump gas production upward and start using as much of that gas as possible to benefit local consumers. Africa can (and should) utilize its gas to produce more electricity; along the way, it can (and should) use its gas to produce propane and butane, both of which can be components of liquified petroleum gas (LPG), a cleaner-burning fuel for use in cooking or home fuel generation.[19]

So yes, I'm recommending that every African state that possesses commercially viable gas reserves move forward with development. I say that if Africa has the gas, it ought to use it to meet its own needs (and export the surplus, assuming there is one)—and it ought to do so on the largest scale possible and as quickly as possible.

Phasing in Renewables Over Time

Prioritizing gas doesn't mean ignoring renewable energy and making climate change a low priority. In other words, it doesn't mean *only* using gas. Instead, it means focusing on gas *first.* It means using gas to pave the way for universal electrification and phasing renewables in over time. It means starting with gas and looking for ways to strike a balance.

It also means acknowledging that renewables have huge potential—and huge limitations.

Solar, wind, and other renewable energy sources do hold the promise of reducing carbon dioxide emissions, but they also have significantly lower power density than fossil fuels.[20] They're more expensive, and the energy they produce is harder to store, since batteries (which are also expensive) have limited capacity. As renewable energy technology improves, we should be able to relegate gas-fired plants to back-up units, useful only for covering the gaps that emerge when the sun doesn't shine on solar panels and the wind doesn't turn turbines.[21] This coverage will become less and less necessary as battery technology improves over time; as it does, gas plants can be phased out.

But this will not be a quick process. It will play out over time, perhaps over several decades. And in the meantime, gas is available, and it is plentiful. It is something African countries can use for themselves, to build themselves up and develop their own economic and industrial capacity, rather than waiting for the developed world to hand down aid. I believe they ought to do it.

I make this statement with a caveat: I believe Africa ought to develop its gas reserves, but with the goal of replacing gas with renewable energy in stages—and doing so carefully, to make certain that the transition does not disrupt the economic gains realized from eliminating energy poverty.

To do this, electricity production from renewable sources should become just as much of a native African industry as the gas-fired electricity production it's meant to replace. African countries should therefore make a careful inventory of their natural resource endowments and capabilities, with the aim of figuring out how they can contribute to the renewable energy value chain. Of course, they can do so on the raw materials end, serving as sources of cobalt, rare earths, and other minerals. But they can also work to develop their capacities in other ways—such as by offering financial incentives to companies that hire local workers for processing and assembly facilities; by making investment deals conditional upon technology transfer and training commitments; or by reserving certain types of service contracts for local companies.

Taking this approach will give Africans multiple ways to be intimately involved in the process of utilizing renewable sources to keep energy poverty at bay because it will ensure that Africans are members of the crews that design, manufacture, assemble, install, commission, connect, supply, service, and manage the parts, equipment, and facilities that make up the continent's new power sources. For a concrete example of this, see the Democratic Republic of Congo's plan to develop its domestic capacity for manufacturing batteries in order to add value to its own production of copper and cobalt.[22] Both of these minerals are used in lithium-ion batteries, which can hold a charge for electric vehicles or store production for a renewable power plant.

If this process goes well, Africa should be able to phase fossil fuels out and replace them with renewable energy—but only with time. It can't make all the changes now, and it can't complete the entire transition immediately. But I believe it can be done, with a little patience and practicality.

This gradualist approach is sure to disappoint representatives of non-governmental organizations (NGOs) and international institutions who have insisted that Africa abandon all fossil fuel development and focus solely on renewables. However, I'm convinced that their disappointment is misplaced—and more than a little unfair.

Many of those who take this perspective hail from countries that have already taken advantage of fossil fuels to move development forward; yet they propose to deny African countries, that currently account for only a tiny fraction of total carbon emissions, the opportunity to do the same. In doing so, they are demonstrating their willingness to consign hundreds of millions of Africans to many more years of energy poverty for the sake of their zero-emissions ideology. This ideology is hardly infallible, as demonstrated by the fact that European energy producers had to resort to burning coal in the winter of 2021-2022—and that was before the Russian invasion of Ukraine sparked even deeper discussions about the energy future of the West.

So let Africa chart its own destiny! Let Africa make its own choices. Let Africans decide whether to use their own resources.

And let Africa start with gas.

Chapter 2

MORE HARM THAN GOOD: HOW A RUSHED ENERGY TRANSITION CAN HURT AFRICA

DESPITE ALL SIGNS pointing to a warming planet, in the fall of 2021, Europeans were bracing for the prospect of a colder-than-normal winter—and its potentially crushing effect on their pocketbooks.

Their concern, however, had less to do with meteorological forecasts than it did with continental gas prices, which by September 2021 had reached multi-level highs. In just the first 9 months of the year, gas contracts pricing at the Dutch Title Transfer (TTF) hub—a virtual trading post for natural gas that provides benchmark pricing in Europe—had risen 250%.[1] By the third week of September, gas futures were trading at USD85.69 (€73.150) per megawatt-hour.[2]

Although this price represented a slight decline from the September 15 record of USD93.31 (€79),[3] it still meant that if the winter months brought anything but the mildest of temperatures, Europeans would be struggling to pay their bills or left shivering in the dark.

The problem was expected to be particularly pronounced in the U.K., where customers pay more for their electricity than anywhere else in Europe. But France and Germany weren't far behind, as their benchmark power contracts doubled in 2021.

The meteoric increase in energy prices was labeled everything from "a perfect storm"[4] to a "cocktail of pretty potent things,"[5] but it came down to this: in an effort to curb the effects of climate change and meet net-zero pledges, the continent began moving away from fossil fuels a couple of years ago. At the same time, they began relying more and more on wind and sun for energy. Even the IEA called for stopping natural gas investments, although it later admitted that natural gas would remain an important component of electricity security, especially in countries where electricity demand varied considerably by season.[6]

The transition toward renewables was all fine and good until the winter of 2020, when unseasonably cold temperatures meant drawing on lower-than-normal gas supplies. That was followed by the summer of 2021, when there wasn't much wind to convert to electricity. Add to that the economic recovery following the initial waves of the pandemic, a hopeful return to some sense of normalcy, but a drain on energy reserves. Suddenly, by the third quarter of 2021, the world's most "enlightened" continent was going so far as to reconsider the use of coal. Yes, coal: the world's most carbon-intensive fuel, which produces nearly twice as many greenhouse gas emissions as natural gas per unit of energy.[7]

The irony should be wasted on no one.

Yet in what seems to be a "do as I say, not as I do" move, this about-face didn't stop European nations from further reducing funding for fossil fuel energy projects, including in poor countries that depend on coal, gas, and oil for state revenues as well as for warmth, light, and cooking.

The question is: why? Even when they see what impact a rushed overreliance on renewables can have on their own economies and citizens, why are richer nations forcing their climate agenda onto struggling and emerging African nations? After all, these countries produce a minuscule fraction of the world's CO_2 emissions, especially in proportion to the size of their populations. Consider this: Approximately 16% of the world's people live in Africa, but they produce less than 3% of cumulative global carbon emissions.

And even if we were to make a meaningful improvement in electrification on the continent—a continuing problem in large part because the hydrocarbons Africa produces have historically been exported to fuel industrial development elsewhere instead of providing domestic power—it wouldn't have too much of an effect on the total.

As Liberia's former Minister of Public Works Gyude Moore and The Breakthrough Institute's Vijaya Ramachandran wrote for *The Hill*, "If the 48 nations in Africa tripled their electricity consumption overnight, relying entirely on natural gas, the resulting carbon emissions would still equal less than 1% of the annual global total."[8] At the same time, the IEA suggests that, if done with the appropriate sources, full electrification will only "add less than 0.2% to CO_2 emissions."[9] Not to mention that several countries are "carbon sinks," meaning their forests reduce more emissions than the country releases into the atmosphere.

As Moore and Ramachandran noted, the evidence shows that a higher-emissions Africa will actually be more resilient to the impacts of climate change. "People with better access to education, health care, and housing are able to cope better with heat waves and typhoons," they wrote. "Roads, hospitals, resilient power grids, early-warning systems, robust food supplies, and other features of modernity buffet societies against natural disasters and other climate risks, even if the energy needed comes partially from fossil fuels. Keeping Africa poor to fight climate change will do nothing to help the people most affected by it."

To some degree, it all comes down to three little letters: ESG.

Too Much of a Good Thing?

ESG stands for environmental, social, and corporate governance. It's more or less what used to be known as "sustainability."

In a nutshell, ESG criteria are non-financial performance indicators that aid investors in determining how well a company is doing as a steward of the Earth and the environment; how well a company treats its employees, suppliers, and customers, as well as the communities in which it operates; and how ethically the company behaves on taxation, political donations, and board diversity, among other issues.[10]

ESG certainly sounds like a good thing, and it is. In a sense, it's a watchdog that reminds companies of their responsibility to help society overcome its shared social, environmental, and economic challenges. ESG encourages businesses to recognize and change the negative consequences of their operations and decisions. ESG is a way for companies to make the world a better place, if you will, and presumably, not do any harm in the process.

It's hard to argue with that.

In recent years, ESG investing has become increasingly popular. It may not be the darling of Wall Street and London (that title probably goes to SPACs), but it's a pretty close runner-up.

No longer concerned solely with the profit motive, ESG investors put their money in companies that score high on scales largely determined by ESG research firms. Businesses in energy and other heavy industries that often don't fare well by ESG measures are under pressure from shareholders to meet stricter ESG standards.

The State of African Energy 2022, published by the African Energy Chamber, described the efforts major oil and gas companies are making to meet carbon-neutral goals: increasing efficiency measures, eliminating flaring, optimizing operations to minimize their carbon footprint, investing in renewables and lower carbon intensity liquified natural gas (LNG) projects, and, importantly, selling carbon-intensive crude oil assets to national oil companies (NOCs) and international

national oil companies (INOCS). "With the European governments imposing binding emission targets, and the European Investment Bank announcing an end to investment in African oil and gas, the majors follow their footsteps," the report said.

And while that enables African companies such as Algeria's Sonatrach, Libya's National Oil Company, and Nigeria's Nigerian National Petroleum Corporation to gain a greater share of domestic output,[11] the overall picture remains gloomy.

Investors everywhere are still leaving oil and gas companies in droves, bolting before businesses even have a chance to change their ways.

In the U.S., for example, the total value of energy companies listed on the S&P 500 is just 5% of the total index. A decade ago, it was more than twice that, at 11%.[12]

And the noose is tightening even more, and not just from investors. In what was called a "precedent-setting judgment,"[13] in May 2020, a Netherlands court ordered Royal Dutch Shell to cut its CO_2 emissions by 45% compared to 2019. The verdict held Shell responsible not only for its own emissions but also for those produced by its suppliers. According to Friends of the Earth, which brought the lawsuit, this was the first time "a company has been legally obliged to align its policies with the Paris Climate Accords."[14]

It should be noted that Shell has a record of emissions reductions, has pledged to become net-zero by 2050, and has broken new scientific ground when it comes to alternative and low-carbon fuels.[15]

Yet as Dr. Barry Po, EVP & Chief Marketing Officer of Artificial Intelligence (AI) company mCloud Technologies, wrote for *Forbes*, even companies that "claim to have reduced a million tonnes of CO_2 equivalent emissions will need to be ready for the Deloittes and KPMGs of the world to ask how those numbers were determined.[16] An ESG reckoning has arrived for the oil and gas industry."

Unintended Results, Real Solutions

What does an ESG-driven reduction in oil and gas investment mean for Africa?

With less capital to support them—and, of course, their own carbon reduction and climate goals in mind—energy companies are moving forward cautiously in Africa, if at all. Shell, for example, has put off making a final investment decision in the huge Bonga Southwest/Aparo oil field in the southwest of Nigeria. ExxonMobil has exited offshore Ghana. Equinor dropped its extensive exploration acreage offshore South Africa.

Other companies are still in the game but switching the balance in their portfolios to favor lower-emission natural gas development. Today, natural gas represents 22% of total hydrocarbon production in Sub-Saharan Africa and that number is expected to climb to 30% in less than five years.[17] However, Conor Ward, oil and gas analyst at GlobalData, cautions that the forecast depends on final investment decisions by exploration and production (E&P) companies—which ultimately rests on how much money investors are putting into those businesses.

The move toward natural gas has geographic repercussions, Ward said. Investment is steering away from developed countries like Nigeria and toward what he called "frontier countries": Mauritania, Senegal, Mozambique, and Uganda. Those nations are offering more attractive fiscal terms that lower the cost of production and their reservoirs are still largely undeveloped.

Still, climate concerns and protests haven't made it easy to get Uganda's first oil projects off the ground.

Consider the East African Crude Oil Pipeline (EACOP), a project of TotalEnergies (formerly known as Total) that will connect oil fields in Western Uganda to an offshore loading terminal in Tanga on the Tanzanian coast. The pipeline project has repeatedly been challenged by global environmental groups who rattle off a list of the ways they believe a Ugandan pipeline will adversely affect local communities, water supplies, and biodiversity. According to S&P Global Platts, "more than 260 charities and organizations from 49 countries called on banks not to participate in loans to fund the construction of the line."[18]

Despite the tremendous level of opposition, TotalEnergies has soldiered on with its plans for the $3.5 billion heated oil pipeline as well as with the Tilenga oil exploration project. Tilenga, which will also include a crude oil processing plant and additional infrastructure, is in the Buliisa and Nwoya districts of Uganda, not far from Lake Albert and Murchison Falls National Park.

Production and export are expected to begin in 2024/2025.[19]

To be able to conduct activities near an area that's environmentally sensitive, TotalEnergies E&P Uganda has committed publicly to limiting their effects. The company's mitigation plan resulted from a four-year environmental and societal impact study that included consulting more than 10,000 people.

As a first step, TotalEnergies says it will voluntarily limit the Tilenga project's footprint within the park; although it has permits that cover nearly 10% of the park, it has pledged to keep development to less than 1% of its surface area. Undeveloped regions will be voluntarily relinquished without delay, a company statement said.[20]

That's just part of the plan, though.

TotalEnergies intends to improve the area, not exploit it. In exchange for the privilege of developing Uganda's natural resources, TotalEnergies says it will:

- Provide support to increase the number of rangers in Murchison Falls Park by 50%.
- Work with the Uganda Wildlife Authority to reintroduce the black rhinoceros in Uganda.
- Create as many as 6,000 jobs, with 70% of them going to Ugandans.[21]

In short, TotalEnergies is publicly committed to leaving the area in better condition than it was before the company began development, including employing 4,200 Ugandans in the process. But let's consider what might have happened had TotalEnergies capitulated to protestors' demands and not moved forward with EACOP and Tilenga.

We'll begin with a bit of historical context.

According to the World Bank, the Ugandan economy was humming along nicely in the 1990s and early 2000s, and poverty was on its way down.[22] Uganda was so successful at implementing economic reform, in fact, that in 1997, the nation was one of only a few to receive international debt relief, and it has continued to qualify for debt relief since.[23]

Since 2011, however, Uganda's economy has slowed, making it more difficult to fight poverty. The pandemic didn't help matters, of course; in 2020, GDP was only 2.9%, compared to 6.8% in 2019.[24]

Experts say that Uganda needs two things to grow: economic diversity and more exports. The EACOP and Tilenga projects provide both.

Without the proposed economic shot in the arm (and glimmer of hope) provided by oil development, there's no question that Uganda's stall will be more protracted, the way out even rougher, and more people plunged back into economic despair.

Don't just take my word for it, though. Even the World Bank acknowledges that oil production is required to get Uganda back on track.

Although the Group Bank says there are still constraints to social and economic transformation, the bank has asserted that "the start-up of oil production and revenues has the potential to accelerate growth by addressing some of the constraints… Actual revenues from oil will depend on levels of production and international oil, with significant impact on government revenue, exports, and investments."[25]

In other words, Uganda *needs* oil development to avoid large-scale job loss, bankruptcies among Ugandan companies, austerity measures, and possibly even instability. It *needs* those 4,200 local jobs. What it doesn't need are outsiders concerned about protecting the continent from itself and trying to limit economic advancement and disrupt African progress in the name of climate change.

I get it, I really do. Everyone wants the world to be a cleaner, healthier, and safer place, not just for today but for future generations. The problem is that when climate activists and ESG investors look at oil and gas companies, just one word comes to mind: dirty. What they're forgetting—or probably overlooking—is how a vibrant and greener energy industry can be a panacea for impoverished nations. What they refuse to see in their self-righteous words and actions is that when oil and gas majors reduce their presence on the continent and African oil and gas investment dries up, so does opportunity for Africans.

Banning energy projects in countries that rely on coal, gas, and oil for revenue has the potential to entrench them in poverty and disarray.

That last statement doesn't affect just a nation or two, by the way.

A Dangerous Multiplier Effect and "Doom Loops": When Hopelessness Takes Root

Two-thirds of developing countries, many of them in Africa, depend on commodity exports, including oil and gas. Five of the world's top thirty oil-producing countries are in Africa. As proven oil and gas reserves have grown in Africa over the past decades, so has export income. In 2016, oil and gas represented more than 57% of the continent's total export earnings.[26] More to the point, nations' oil exports account for up to 90% of revenues, according to Deloitte,[27] which reminds us that in 2000, *The Economist* called the continent "hopeless," only to do a 180-degree revision a decade later, thanks in part to the strength of the growing African energy industry.

In country after country, African leaders are putting oil and gas revenues to work for ordinary citizens, using at least some of those dollars to prop up social programs while also keeping the government running smoothly. And while it's easy to argue that the money could be spent more efficiently—some have even proposed it become the basis of a form of universal income—the fact is, the fewer the energy-related dollars, the worse the situation.

We've seen the fallout from low energy prices. Spending cuts, currency weakness, higher inflation, and lower investment all accompanied the collapse of oil prices in 2014, and it's not hard to imagine how declining support for African oil and gas by investors from far richer nations could multiply those woes.

In fact, you don't have to imagine it. The risk consultants at Verisk Maplecroft have laid out the future quite plainly.

In a 2021 report, these risk consultants said that as oil producers pull back because of the energy transition or if fossil fuel activity is abandoned altogether, political instability is likely to result. Algeria, Chad, and Nigeria are among the first countries that will feel the fallout.

According to the report, "The most vulnerable countries are higher-cost producers that are heavily dependent on oil for revenues, have lower capacity to diversify and are less politically stable." For the countries that are hit the hardest, the result could be "doom loops of shrinking hydrocarbon revenues, political turmoil, and failed attempts to revive flatlining non-oil sectors."[28] Mozambique provides a real-life, too-close-for-comfort example of what that might look like. The country, long plagued by poverty—it's sixth on the list of the world's poorest nations—and political corruption has become a breeding ground for extremism in recent years. Since 2017, an armed opposition group calling itself Al-Shabaab, which translates to "the youth" in

Arabic, has burned villages to the ground and killed thousands of civilians, often by beheading. (The group is not tied to the similarly named al-Shabaab in Somalia.) It's estimated that half a million people have fled Mozambique for fear of being next.[29] Amnesty International says that when they interviewed displaced young women and girls, many said they left Mozambique to avoid being abducted, detained, raped, or forced to marry Al-Shabaab fighters.

At the root of the violence is anger, despair, and hopelessness. It is believed that Al-Shabaab gets most of its support from young, unemployed people in Cabo Delgado, in the country's northern region, who feel that the government has kept them from benefiting from the province's ruby and natural gas industries.[30]

Natural gas was found offshore in Cabo Delgado in 2010 and the area may hold as much as 125 trillion cubic feet (tcf) of gas,[31] reserves so large they could eventually put Mozambique into the top tier of global producers of liquified natural gas (LNG).[32] Western, Chinese, Malaysian, and Japanese energy majors have all set up shop in the area and are developing LNG projects valued at USD60 billion that were originally set to come online in 2024.

The huge market for LNG—global demand increased to 360 million tons in 2020, with China and India as the biggest users—provides Mozambique with a promising path to economic growth and diversification. Exploiting the vast swath of natural gas is likely to expand access to electricity and generate revenue for Mozambique, a low-income (and low-carbon-emitting) country. By strengthening the economy, LNG could also facilitate an end to the local conflict.

However, at this point, the violence is getting in the way.

In April 2021, TotalEnergies suspended operations in the area after a deadly raid on a neighboring town. At USD20 billion, the TotalEnergies gas liquefication plant is the largest foreign investment in Africa. In August, TotalEnergies announced the project might resume in 18 months if African armies sent to assist Mozambican forces can stop the insurgency.

Unfortunately, the bedlam isn't limited to Mozambique. In 2021, there were two coup d'états, in Mali and one in Guinea-Conakry, along with an attempted coup in Sudan, reversing the decades-long period of democratization Africa had enjoyed. They are the result of economic hardship—a condition that's only going to get worse if African oil and gas jobs are lost due to net-zero ambitions and the lack of industry investment. The *State of African Energy 2022* report says that as demand gradually shifts toward renewables and low carbon-emitting sources, employment in the oil and gas industry in Africa will shrink on average 12.5% per year between now and 2030.[33]

There's a lot to unpack here: poverty, disaffected youth, corruption, lawlessness, brutality. The situation in Mozambique proves that violence never solves anything. The tremendous loss of life is almost too much to bear, and it's perpetuating the cycle of poverty and desperation that feeds terrorist activity to begin with.

There is hope, though, that once calm and confidence are restored in Cabo Delgado, once the chaos is put to rest, the country will be able to avoid future instability, tap into its natural resources, and create an economy that works for everyone.

Why Single Out Africa?

We've covered a lot of ground in this chapter, but let's end where we started: with Europeans getting ready to fire up their coal plants. No one is shaming them or tallying up the extra emissions that might result. No one is calling for a blanket ban on infrastructure investment. They are accepting the reality of the situation, doing what they have to now, and planning to do better in the future.

Still, the rug is being pulled out from under oil and gas investments in Africa.

What I'd like to ask these European nations is how much they think their actions will affect global greenhouse gas emissions.

Truth be told, it's unlikely to matter much at all.

Remember: Even with 16% of the world's population, the entire continent of Africa produces less than 3% of annual global emissions.

As James Murombedzi, coordinator of the African Climate Policy Centre, said, it is not reasonable for the rich countries that produce the vast majority of global emissions to expect or insist that Africa abandon its economic opportunities to do more than their fair share to reduce emissions. A better idea, says Vijaya Ramachandran, would be for the European Union, United States, and World Bank to adopt funding criteria that consider economic growth alongside climate impact.[34]

As for me, I believe that a rushed transition to renewables and a blanket ban on African oil and gas will force hundreds of businesses into bankruptcy, plunge billions of people into irrevocable poverty, and destabilize more governments. We Africans need the time to capitalize on our oil and gas resources while we build opportunities in renewables. We want our voices to be heard, our opinions to be counted.

Otherwise, we'll find that, just like tight natural gas sources during a winter freeze, the costs are simply too high.

Chapter 3

THE REALITIES OF ENERGY POVERTY IN AFRICA

When Boston physician Khameer Kidia called his mother in Zimbabwe to tell her he'd soon be getting vaccinated against COVID-19, his good news was overshadowed by guilt. Although his mother was happy for him, she had a much different outlook on her future.

Kidia wrote about their conversation in a December 2020 article for *Slate* magazine. "'I'm not sure a vaccine will work in Zimbabwe,' my mother says to me. It's the middle of a hot summer in Zimbabwe. There are power outages for hours each day. The food in my mother's freezer spoils in the heat. Even if by some miracle the government could pay for the vaccine, how will they keep it cool? How will they deliver it to rural areas where most Zimbabweans live and where, even in many clinics, reliable refrigeration is a luxury?"[1]

Even before the pandemic, lack of electricity has been a pervasive problem in the continent. More than 600 million people in Sub-Saharan Africa have no access to electricity. Hundreds of millions more have unreliable or limited power at best; they lack the energy to fuel things that people in other parts of the world take for granted—clean and safe drinking water, social services, and access to information. In other words, they live in "energy poverty."

Energy poverty is commonly measured as very low average kilowatt hours (kWh) consumption per person. But this is not really an accurate assessment. For a more complete picture, we need to consider the extent to which a lack of energy prevents the fulfillment of basic needs. And this is really a universal consideration: we all need electricity to run a household—no matter where we live. Of course, those needs might differ from place to place: an affluent community might include luxuries like heating, air conditioning, refrigeration, and other household appliances, while the energy needs of a developing neighborhood might be somewhat simpler.

According to Energypedia, "More than 80% of the energy expenditures of poor households are spent to satisfy the energy needs for cooking, lighting, information, and communication media."[2] Energypedia uses five criteria to determine if energy poverty exists on a household level:

1. All members of the household have the opportunity to work or relax under sufficient light for at least five hours each day.
2. The household has the means to prepare typical hot meals for everyone in the home.
3. The household can use a telephone whenever needed and a radio/TV for several hours each day.
4. The regular household income can cover the expense of the energy.
5. Hardware, fixtures, and energy sources are safe and do not shed toxic particulates.

But energy poverty goes beyond individuals lacking access to energy. Energy poverty has wide-reaching implications that extend well beyond a home, a neighborhood, or a town, from economic stagnation to lack of medical care. The prosperity and development status of whole nations—and entire regions—are closely correlated to the type and extent of access to energy of its inhabitants.

In Africa, we have some of our smartest minds working on alleviating energy poverty, and I believe that we are on the precipice of exciting breakthroughs. But while we watch with bated breath to see what new technologies will emerge, we must not scrap all of our efforts in the petroleum industry. We must not abandon the successes we've experienced thanks to oil and natural gas, and we must remember that a solution to energy poverty is a *comprehensive* strategy whose achievement depends upon the incorporation of all sources of viable energy at our disposal. Even as we look to the future of new or renewable resources, we shouldn't abandon our present use of natural gas, which gives people across Africa access to more reliable energy.

Reasons Behind Energy Poverty

The problem of energy poverty is particularly rampant in Africa. Consider this: A quarter of the world's population lacks access to energy, and of the 32 countries around the world with an electrification rate of less than 50%, 26 are located in Sub-Saharan Africa. The region's electrification rate average is just 26%, and rural areas average a mere 8%.[3]

And even where electrification rates are high, connected households may not have reliable energy because of voltage fluctuations or frequent outages. Access is often determined by the quality and consistency of the local grid.

This is why "rolling blackouts" are a common occurrence in many African countries. In South Africa, for example, the single electricity utility, Eskom, has blamed the power system's vulnerability to unplanned breakdowns on years of inadequate maintenance. To compensate—and ideally avoid a nationwide outage—the utility maintains a schedule of controlled blackouts along the national power grid. In this method of power rationing, sections of the grid are shut down for a specified timeframe of up to 12 hours. At the same time, however, the utility has acknowledged that unplanned outages are still a reality.[4]

Access to energy isn't the only barrier, however. Many people who have access are limited by the cost-prohibitive nature of the energy available to them. Statistics from the World Bank show that consumers in many countries in Sub-Saharan Africa pay as much as 20 to 50 cents per kilowatt-hour, compared with the global average of about 10 cents.[5]

But there are other issues as well: inadequate energy efficiency in the form of poor insulation or inefficient appliances adds to the cost. And, as a 2017 Organisation for Economic Co-operation and Development (OECD) report described, on a global scale, there are many more households experiencing fuel poverty than there are households below the poverty line: "Findings show a clear link not only with low disposable income, but also poor housing conditions that lead to low energy efficiency, which means that fuel poverty in some countries concerns a much larger group than those on very low incomes."[6]

Add COVID-19 to the equation, and the situation only worsens. From the economic devastation caused by lockdown measures to governments shifting their focus and capital expenditures to the health sector, the pandemic has been erasing progress made in recent years to increase access to energy. In fact, the International Energy Agency (IEA) estimates that more than 30 million people (worldwide) who had power retreated back into energy poverty at the end of 2020 because of affordability challenges.

Painful Impacts

Energy poverty can cause dangerous choices. Without access to modern-day energy sources, the livelihood, well-being, and health of millions of people around the world are diminished by the harmful energy options they have at their disposal.

In large swaths of Sub-Saharan Africa, much of the population cooks over stoves burning solid fuels, including biomass fuels such as wood, animal waste, agricultural residues, and coal. Each year, millions of deaths worldwide—estimates range from 1.5[7] to 4 million,[8] most of them women and children—are attributed to illnesses and accidents caused by cooking with these harmful fuels.

Meanwhile, it's estimated that 58.3 million households[9] around the globe rely on kerosene for indoor lighting. Kerosene fumes are linked to lung cancer, pneumonia, heart disease, and other health problems. Kerosene lamps are also a leading cause of many house fires.

Energy poverty also prevents access to clean water for some 45% of Sub-Saharan Africa.[10] Without the energy-based technologies for purification and distribution of water supplies, Africans are needlessly subjected to disease, poor hygiene, and inadequate sanitation. The World Health Organization reported in 2022 that approximately 829,000 people in low- and middle-income countries die as a result of inadequate water, sanitation, and hygiene every year.[11]

At the time of this writing, only a quarter of Sub-Saharan African health care facilities have reliable power. This means that hospitals and care providers are often being tasked to treat patients without modern equipment (or sometimes even lights). According to a study of 33 hospitals in 10 countries around the globe, unreliable power was the single most common cause of medical equipment failure.

"But electricity access affects human capital, too," a Brookings Institute blog pointed out. "Health care workers prefer living in villages with access to electricity, and this in turn reduces absenteeism. Research on job satisfaction among health care workers finds that electricity deficits turn simple tasks like putting in an IV line into frustrating challenges. In the study, nurses and doctors expressed fear for their own lives as they handled blood samples and other potentially contaminated fluids when working in the dark."

Stunting Economic Growth

So, how can we begin to address such a dire threat to Africans' well-being? The solution must begin by breaking the challenging economic cycle we find ourselves in. As Sri Mulyani Indrawati, the World Bank's former Managing Director and COO, pointed out, "Inclusive economic growth is the single most effective means of reducing poverty and boosting prosperity. Yet most economic activity is impossible without adequate, reliable and competitively priced modern energy."[12]

Put another way: Access to energy services can actually stimulate greater social, economic, and environmental development.

Ensuring energy access expands career options, particularly in rural locations. Estimates range, but 60% to 90% of the population in low-income countries likely depends on agriculture, fishing, and forestry as their main source of income.[13] In many cases, these limited options are a result of energy poverty; industrial, professional, and services-related industries are difficult to impossible without reliable power. And agriculture, fishing, and forestry can provide sustainable wealth, but they rely heavily on cooperation from Mother Nature. Without economic diversification, communities are much more susceptible to the chains of long-term poverty.

Unfortunately, power interruptions across Africa are frequent, prolonged, and unpredictable. All of this combines for an adverse effect on the economy, limiting outputs in productive sectors and discouraging new investors or firms from locating to the area. The result? Fewer new jobs are being created, household incomes are lower, and less revenue is flowing into government coffers.

In short, because of this limited and unreliable energy supply, the chips are essentially stacked against African businesses from the start.

One recent study conducted for the Center for Global Development found that African firms are about 20% to 24% smaller than firms in other comparable nations due to a number of factors, chief among them a consistent lack of reliable electricity.[14] In this environment, even the most productive firms cannot grow.

So what's a company to do?

Short-Term Fixes as a Bridge to Long-Term Solutions

Building more power plants is one method to meet this demand for reliable power. Before this can happen, though, we must make some significant changes: Governments should remove subsidies and introduce optimal tariffs to promote investments in new grids—and attract international and private investors. And lawmakers should introduce incentive regulations that encourage the participation of the private sector in the generation of electricity.

But changes like this don't happen overnight. While lawmakers deliberate the issue of energy poverty, those millions of Africans without access to power continue to suffer. We can't keep waiting for external forces to drive our growth. There are some immediate, short-term fixes we can adopt across the continent that certainly aren't solutions for energy poverty but can become stepping stones that pave the way toward energy security for all.

To cope with prevalent power outages, businesses, schools, and medical facilities across Africa resort to generators to keep operations running in the short term. Generator ownership and the share of electricity coming from self-generation correlates to firm size: Larger companies are more likely to be able to afford this option.

One solution to outages could be shared generators with small enterprises or communities pooling together for backup power. If governments formalize systems of generator-sharing, particularly in "industrial parks," this might boost access to generator-powered electricity among businesses of all sizes.

But self-generation isn't really a viable long-term solution. It affects investment decisions, the cost of production, and firm location. This option is more expensive than energy supplied by a public utility. It's also less effective, meaning companies might not be able to operate at their full capacity, thus reducing product quantity and quality, halting production, and delaying order delivery. When businesses focus investments on self-generation, it forces them to channel their much-needed capital into financing less productive equipment.

Until we connect more people to the grid, there are other steps we can take now to protect communities and give businesses a fighting chance of thriving.

For one, we must continue efforts to educate people about the dangers of cooking indoors and performing household chores with biomass like firewood, charcoal, and dung. We must encourage them to choose liquid or gas fuels such as ethanol or natural gas while waiting for electrification via a national grid.

We must then identify and promote energy options that are appropriate for specific communities and populations, meaning they must be affordable, available, and sustainable for the basic energy services needed at that level. And the investment costs must be low to ensure compliance. At a minimum, the options offered must improve living conditions for a large number of people almost immediately.

We must also accept that deficiencies and inequities in the quantity and quality of energy supply will differ from region to region. For example, because it's easier and faster to expand the electricity grid where the infrastructure is already in place, urban populations will benefit more than rural locations. Similarly, promoting the use of liquified petroleum gas (LPG) will be much more effective in densely populated areas that already have an established system of fuel distribution. Conversely, attempting to extend grid-based electricity or commercial fuel-based energy into a rural community is often not feasible, either economically or technologically.

Certainly, we're seeing small-scale renewables projects making inroads. So-called "mini-grids," for example, are becoming popular alternative energy services in Sub-Saharan Africa that can be increased or expanded as the demand grows. These solar (or solar-diesel hybrid) "plug-and-play" systems have been touted as an economical way to deliver reliable, on-demand electricity to millions across Africa. Because these are relatively quick to put up and don't rely on a central system, they can begin to generate and deliver power without communities waiting for years to be added to the grid. But, as we'll explore further, mini-grids come with their own challenges, from consumers' struggles to pay for service to the challenges of conducting maintenance and repairs in remote areas.[15]

Small-scale solar power generation is another area with potential. Portable solar home systems and appliances can bring much-needed energy to households, while simple solar distillation techniques can improve water purity. As these technologies continue to improve, the benefits to whole communities can be seen.

And while we're conquering the immediate energy needs of the African people, we can begin to address the long-term solutions to sustained and sustainable power.

Despite progress, current and planned efforts to provide access to modern energy services barely outpace population growth. By 2030, the International Energy Agency (IEA) estimates that 530 million people will still lack access to electricity, and twice as many still won't have access to clean cooking sources. By 2040, 90% of those without electricity access and 50% of those without clean cooking will be living in Africa.[16]

These statistics are shameful.

Let me be absolutely clear: I am all for renewables. But *not* to the detriment of African people—and not at the cost of Africa's petroleum industry. We can benefit from continuing oil and natural gas operations while we are developing the technologies to support renewables. At present, we don't have the sustainable energy solutions to meet Africa's power needs. It is impractical and, frankly, inhumane to force Africa to abandon the readily available natural resources we're already successfully producing, just to further a green agenda.

So, if the current and planned efforts aren't enough, what should our next steps be?

We must attack the problem by understanding that every region, country, community, and even household is unique. Of course, we need quantitative targets as a starting point to facilitate the management, monitoring, and verification of implementing widespread energy access. These targets should consider the availability, affordability, equitability, suitability, and sustainability of proposed energy solutions. But we must ensure that our leaders (*and* international investors) respect and support the highly specific local objectives of the people, who must be included in all stages of the development process.

In the long run, we need political reforms in the energy sector that will foster investment in forms of energy that will benefit the poor. So far, though, most investments only an indirect effect on access rates. We need investment policies that directly target access among households lacking modern energy services. We must work to build up local capacity via methods that can be replicated to ensure long-term sustainability. We must make sure that Africa's energy industry is a multi-pronged partnership that includes *all* the natural resources at our disposal.

And, perhaps most importantly, we must work towards these goals on our own terms, and we must aim to do this without relying on handouts or help from (mostly) well-meaning outsiders.

Chapter 4

FOREIGN AID TO AFRICA: IT'S TIME TO END THIS HARMFUL CYCLE

IN 2017, ANTHONY Leddin was training Ethiopian dairy farmers to raise forage crops for their cattle when he realized something wasn't quite right. Leddin, an Australian plant breeder, is a volunteer for a non-profit organization called Breeders without Borders. Here he was, trying to help African farmers become self-sufficient and alleviate famine in their country, yet the free seeds they were using to accomplish that goal had been shipped in from Western countries, not harvested or even packaged in Africa. In other words, the seed supply chain was missing an important link: It didn't start in Africa, or even go through Africa until the seed landed in the field.

African seed growers, who could have benefited from the opportunity to help their fellow Africans, were essentially excluded from the supply chain. Not only that, seed flooding in from overseas markets would eventually lower the prices local growers could charge. The solution was unsustainable at best, wrong-headed at the very least, Leddin realized.

“I thought: Why haven’t seed companies established themselves in Ethiopia to sell seed?” Leddin wrote in an article for *Seed World*.[1] “When a famine strikes, plenty of seed is given away to people as handouts. This makes it impossible for a company to try and sell seed when it is being given out for free.”

The impulse to help Africa is not new, and it is often well-intentioned. It’s extremely difficult for Westerners like Leddin to see television coverage of starving babies and not feel compelled to offer aid. That’s one reason Africa has long been the recipient of food and monetary aid.

The problem is that Africa has been receiving aid for nearly six decades, and what good has it done? Unfortunately, the history of aid to Africa is paved with failure.

It may sound counterintuitive, but rather than being a salve to soothe the suffering, financial and other forms of aid, even food donations, can instead fuel the flames of hardship. And when assistance is provided on a national or regional scale, it can negatively impact entire populations.

In fact, foreign aid can create poverty rather than ameliorate it, through “economic institutions that systematically block the incentives and opportunities of poor people to make things better for themselves, their neighbours, and their country,” U.K.-based international lawyer Jonathan Lea wrote in a 2015 article for his firm’s website.[2]

And now, the international community is talking about aid for African countries again, this time as a substitute for oil and gas activity. The argument is that Africa should keep all of its petroleum resources in the ground to minimize greenhouse gas emissions and prevent further climate change. Developed nations could encourage this by compensating Africa for that sacrifice, and Africa could utilize that money to develop other opportunities.

This is a horrible idea. Offering Africa aid packages to halt oil and gas operations will do much more harm than good.

There's no denying climate change. And I'm all for exploring every possible avenue toward renewable energy sources. But I caution that the rush to push through green initiatives may cause irreparable damage to the most vulnerable African communities. The fight against climate change is not at odds with creating economic dividends, and investing in the means to provide universal energy today—with the resources currently at our disposal—will generate revenues to develop widespread clean and accessible energy tomorrow. This is the path Africa should take, not a new chapter of aid and dependence.

Feeding the Problem

The socioeconomic backlash of aid often is a cycle. First, a shortage becomes present in one or more countries. International donors, looking to be helpful, send aid money, food, and/or supplies to those countries. Then local farmers, companies, or industries, unable to compete against "free" products, go out of business. As these local suppliers are no longer around to provide for the needs of their people, the local shortage grows… and the vicious circle starts over again.

This is even true for food aid, which at face value, seems like a logical way to meet vulnerable populations' most basic need: sustenance.

But it doesn't work that way.

Harvard University economists Nathan Nunn and Nancy Qian have named three detrimental impacts of food aid[3]:

1. Increased conflict and violence. Because the recipient governments dictate how donations are used, governmental control becomes more lucrative. Nunn and Qian blame food aid to Somalia in the early 1990s, at least in part, as a cause of prolonged civil war because instead of feeding their people, the government traded the food for money and weapons. Similarly, Nunn and Qian said, food aid in Haiti has caused increasing gang violence, with armed guerrillas seizing deliveries and selling the food on the black market.

2. Improper food distribution. Political motives and the personal interests of recipient governments often preclude the distribution of food to those who need it most. Nunn and Qian cited a 2010 United Nations report that claimed that up to half of the food aid donated to conflict-ridden and resource-poor areas never reaches those who really need it.

3. Destruction of the local economy. Food aid can force local farmers and vendors in recipient countries out of business as demand for and prices of local food decrease. To assist Haitians after a devastating 7.0 magnitude earthquake in 2010, several international humanitarian groups delivered rice in the hopes of staving off hunger. Unfortunately, flooding the market with free rice drastically reduced the demand for rice sold by local retailers, and vendors subsequently stopped carrying rice altogether, which put many local rice sellers out of business. The Haitian government ultimately beseeched humanitarian groups to cease their food aid deliveries because this "good deed" dismantled an entire local network. What's worse, the crumbling distribution network meant that rice couldn't be adequately dispersed. Hundreds of thousands of Haitians continued to starve, despite plentiful rice reserves.

"There is an emerging recognition of the need for calculated and measured solutions to food shortages. Humanitarian agencies must look beyond the short-term goals to the long-term consequences," Nunn and Qian wrote.

Aid at a Cost

I'm not disparaging foreign entities that are genuinely trying to make a difference in the lives of Africans. But I do have a serious issue with foreign stakeholders who feel that providing humanitarian assistance gives them carte blanche to exert undue influence on domestic decision-making.

And with Africa poised to participate in the worldwide energy transition, my fear is that international agencies and organizations will feel emboldened to dictate Africa's policy. For example, they might demand that we achieve an immediate changeover to renewables, or that we replace our current flow of petroleum revenue with foreign aid—a giant step backward in our move toward energy, economic, and even individual independence. What's more, assistance can never replace the oil and gas industry's ability to create jobs and business opportunities, grow local capacity, open the door to technology sharing, facilitate economic growth, and alleviate energy poverty.

Promoting the Dignity of Work

Through work, Africans achieve the dignity of helping themselves, of being responsible for their own well-being, of improving their own conditions.

The dignity of work goes a long, long way in building a strong, sustainable community. Individuals who feel empowered work together to improve their lot because they believe in their ability to effect change. There is a true pride that comes after a hard day's work—a pride that no one can take away. Contrary to continual handouts that perpetrate dependence, an inclusive workforce—peopled by citizens who feel their contributions are invaluable and appreciated—promotes continued growth.

I'm not alone in my opinion. During a September 2019 United Nations General Assembly,[4] Ghana's president Nana Akufo-Addo appealed to the gathering to work with countries to reject a mindset of dependency, charity, and handouts—instead charting their own path of self-reliance. Even with the most charitable donor countries in place, Akufo-Addo said, "There will never be enough aid to develop Ghana, let alone Africa, to the level we want." It was never the intention, he argues, for aid to become the mechanism to elevate African nations to the status of "developed." Aid must instead function to creatively and effectively leverage domestic resources to finance rapid economic growth and social transformation.

Indeed, Akufo-Addo's plea emphasizes what I've been saying: Africans are not looking for, and are not helped by, handouts.

It reminds me of something Nelson Mandela said in 2005: "Overcoming poverty is not a gesture of charity. It is an act of justice. It is the protection of a fundamental human right: the right to dignity and a decent life. While poverty persists, there is no true freedom… Do not look the other way; do not hesitate. Recognise that the world is hungry for action, not words. Act with courage and vision."[5]

That's why we must continue working for a thriving African oil and gas industry. Yes, each oil- and gas-producing nation is different, but we can reach this goal through many of the proactive measures I've long advocated for:, fine-tuning local policies, encouraging and supporting local start-ups and small businesses, monetizing gas for domestic use, investing in energy infrastructure, and bolstering energy security in African countries.

By moving forward with the hard work of accomplishing these goals, we can create sustainable, far-reaching opportunities for Africans. And, perhaps most importantly, we can provide Africans with the dignity of work—and of working for their own futures.

An Eye to the Future

One thing that makes food aid (and other forms of well-intentioned charity) so problematic is that it focuses on the short term. Most foreign aid provides quick fixes but does little to address the underlying issues that keep African communities and people trapped in a cycle of poverty. This is why it is critical that our efforts focus on empowering people and communities and setting them up for long-term success.

If we want to work towards a future of sustainable opportunities, we would do well to take note of some of the initiatives and outreach programs that are set up to positively impact Africans—now and in the future.

One Acre Fund, for example, works toward a lofty goal of ending chronic hunger in East Africa. "One Acre Fund supplies smallholder farmers with the financing and training they need to grow their way out of hunger and poverty," the group's website (oneacrefund.org) explains. "Instead of giving handouts, we invest in farmers to generate a gain in farm income."

At the core of the organization's work is the understanding that food aid is "at best a temporary solution." As such, the group has developed a much more effective solution: participating farmers receive an investment package that supports their efforts to provide adequate amounts of nutritious food for their families. In addition, the group offers weekly classes on innovative farming practices and facilitates access to harvest markets.

As of 2022, the program has served more than 1.4 million farmers and their families in nine African countries. Rather than simply receiving handouts, these farmers are empowered to ensure their own food security while developing their local economies—which, in turn, creates a sustainable approach to development—all without dependence on foreign agencies. The success of the program is proven in the ability of participants to help themselves thrive: The group's 2021 annual report announced that 95% of the farmers had repaid their One Acre Fund program fees in full, and that for every USD1 invested, the farmer produced USD3.60 in additional income.[6]

Perhaps we can also look to Uganda for a different strategy. Since 1963, the Karamoja region has received aid through the United Nations World Food Programme (WFP).[7] But this long-term assistance has created a dependency that, as Ugandan Parliamentarian Peter

Abraham put it, "destroyed the energy and commitment the people used to have to sustain their livelihood."[8] In response to Abraham and others, in 2011, the WFP initiated a new approach to food insecurity in this region. Only the most vulnerable received free food; other residents received assistance to plant and cultivate their own crops, start small businesses, and become self-sufficient. The WFP launched training programs to help farmers increase productivity and diversify crops in order to enhance nutrition.

Today, WFP supports the efforts of 125,000 small local farms to improve household income by coaching them to reduce post-harvest loss, implement farming methods that enhance productivity and food quality, and leverage infrastructure to improve market access.

A third example of positive outreach is Power Africa (www.usaid.gov/powerafrica), a USAID-led initiative that the Obama Administration launched in 2013 to enhance electricity access through technical assistance, grants, financial risk mitigation tools, loans, and other resources.

The initiative aims to build power generation facilities capable of producing 30,000 megawatts of new power and establish 60 million new power connections by 2030. A sub-initiative supports off-grid electricity access for the many people who live in remote locations. To reach these lofty goals, a main tenet of the program is to attract private local capital for investment in energy infrastructure. Instead of being yet another aid program, Power Africa is a multi-stakeholder partnership among the governments of the U.S., Tanzania, Kenya, Ethiopia, Ghana, Nigeria, and Liberia, as well as American and African private-sector entities. The African Development Bank Group (AfDB) was a key partner in the program design and implementation.

"The billions of dollars available for investment in the energy sector will translate into actual bulbs in people's homes and electricity necessary to grow small businesses if state utilities run efficiently and effectively. The policy reforms will facilitate and enhance cross-border energy markets," former AfDB President Donald Kaberuka said at the launch of the program.[9]

Change the Paradigm: Investment, Not Aid

And it's Kaberuka's words about energy reforms so succinctly explain why these programs succeed.

Let's look at the common theme these success stories share: foreign entities are investing in communities rather than handing out alms, thereby fostering community development rather than promoting cyclical dependence. They are looking to build a future instead of exacerbating poverty today.

If the countries of the West truly want to empower African countries to rise to their full potential, the best strategy is to improve the continent's access to markets. I believe—and I'm not alone in this belief—that foreign investors must consider themselves partners to foster trade rather than saviors doling out aid. Consider some of the insights of the non-profit PovertyCure, which seeks to battle local and global poverty by studying the condition and fostering entrepreneurial opportunities in afflicted areas.

"Poor countries grow economically when they are allowed to compete in the global economy and are linked to networks of productivity and circles of exchange," PovertyCure stated. "Business and entrepreneurship are keys to prosperity and economic growth. Transparent and competitive markets are beneficial to the poor. No market economy will ever be perfectly just, but where these institutions are weak or missing, the poor are especially harmed."

Foreign empowerment for African countries, whatever the form, should work to bolster all aspects of an economy:

- ***Secure fiscal and monetary policies:*** Mobilizing domestic resources to address illicit capital flows.

- ***Good governance:*** Transparency to strengthen judiciary systems and free them of rampant corruption.
- ***Increasing intra-African trade:*** Searching out investment opportunities in financing sustainable development.
- ***Local content:*** Training for locals in-country or in the region.
- ***Job creation:*** Recognizing that Indigenous workers are key agents of change, and that reaching development targets can only happen with their active participation.
- ***Education:*** Improved quality and access for all learners.
- ***Infrastructure:*** Prioritizing and expanding innovation, technology transfer, and digitalization.

African nations cannot be passive recipients of this assistance. African governments need to take this opportunity to scale up policies that spur democracy, creating enabling environment that will build prosperity in Africa through concrete priorities such as job creation, regional integration, and economic engagement. It is critical for African leadership to continually enact pro-growth legislation that incentivizes economic development and provides confidence to all investors.

Africa is fully capable of building a better future, ending energy poverty, strengthening our economies, and improving the lives of everyday Africans. If we're smart about it—and we work together with purpose—our oil and gas resources can help us get there. And when it comes to specialized, highly technical industries like oil and gas, it should absolutely be African companies that take the lead in the development of the continent's oil and gas industry.

Investing in "The Now"

Just as successful food programs focus on building a better foundation on which to grow, sustainable energy development in the future requires investing in the present. Energy poverty, to which I'm devoting a chapter to this topic in this book, is a real condition that afflicts millions of Africans today and threatens the continent's ability to move forward into tomorrow. Our first step must be to eradicate this affliction.

How?

Let's not erase the massive successes to date or erode the amazing potential of the oil and gas industry across Africa.

In many ways, the African energy endeavor is akin to the American shale revolution. Its tremendous success hinged on the work of many small local businesses willing to take a chance on a venture into uncharted territory. The hard work and dedication of those pioneers significantly boosted production and pushed the United States to the top position among the world's crude oil producers.

The same thing can happen in Africa. When—not if—we can amass more entrepreneurs willing to seize opportunities, make mistakes, learn and grow the industry. When—not if—we elect government leaders who do their part to create a welcoming environment for foreign investors. When—not if—our decision-makers establish and promote local content policies that result in opportunities for business partnerships, quality jobs, and learning opportunities for Africans. When—not if—we grow strong enough to refuse other outside organizations to interfere with our natural resources.

Indeed, the PovertyCure contends that oil and gas investment and environmental concerns are not mutually exclusive: "Good economic development is sustainable and should be environmentally sensitive. Economic progress is the fastest path to an economically and environmentally sustainable future."

To be successful, an energy transition in Africa must continue to include the strength of its oil and gas efforts. The industries of petroleum-producing countries must be encouraged to continue providing opportunities for both indigenous companies and foreign ones.

Indigenous entities offer jobs to local workers and gain the experience needed to lead their own ventures, while foreign investors learn the local customs and comply with local content laws as they invest in the communities where they work to promote their businesses on a global scale.

Investments in oil and gas operations are ultimately investments in people: bolstering infrastructures, improving living conditions, providing education programs, and offering good jobs. All of this empowers communities to make better lives for themselves.

African countries already have a pathway to a better future. Pushing for seemingly quick fixes, whether they come in the form of aid or a one-size-fits-all energy transition policy, will do more harm than good.

Chapter 5

OIL AND GAS DISCOVERIES WE CAN'T IGNORE

Namibia and Botswana could be on the brink of a promising new reality.

In late 2020, petroleum exploration and production activity in these countries roared into overdrive as Vancouver-based Reconnaissance Energy Africa (ReconAfrica) began exploratory drilling in Northeast Namibia and Northwest Botswana. By April 2021, ReconAfrica released preliminary data from its exploratory wells that confirmed an active petroleum system in Namibia's Kavango Sedimentary Basin, described by some as the "largest play of the decade."

"Analysts estimate the company could generate 120 billion barrels of petroleum on just 12% of this footprint, potentially outperforming the oil-rich Permian Basin in Texas if exploration continues," Petro Online reported at the time.[1]

Since then, ReconAfrica has signed a petroleum agreement with the Namibian state oil company NAMCOR. Recon now owns a 90% interest in the Kavango Basin, and NAMCOR holds the remaining 10% interest. Based on their commercial success, the agreement entitles Recon to a 25-year production license over any commercial discovery.

These developments could be huge for the people of Namibia, as well as Botswana. Exploration and production activities could result in well-paying jobs for local populations, along with the opportunities for building local capacity and technology-sharing that come with the presence of international oil companies. Also critically important, domestic oil and gas production can help Namibia reduce energy poverty through gas-to-power programs.

Of course, if you listen to the remarks of environmental organizations and non-governmental organizations (NGOs), you'd think the discoveries in Namibia and Botswana are nothing short of tragic. You'd think that Namibia and Botswana's people and ecosystems are being jeopardized by the Kavango E&P program—and that petroleum production there will nudge the world toward climate disaster. ReconAfrica has been accused of failing to conduct sufficient environmental impact studies, and environmentalists say their drilling threatens vital water systems as well as Botswana's Okavango Delta, home to a wealth of birds and animals.

The overall implication is that the government and communities in Namibia and Botswana care nothing about their ecosystems and will gladly put them at risk for the chance to take in oil money. That's simply not true. And the argument that drilling in the basin will nudge African nations, and the world, into climate-related disaster[2] is unreasonable, too, especially when you consider the widespread reliance on fossil fuels around the globe that is really driving climate change.

I agree with the comments made by Tom Alweendo, Namibia's Minister of Mines and Energy, about moving forward with the E&P activity: "Any volume of oil that is commercially viable will mean a lot to our economy, not only in terms of employment, but income that would come into the treasury," Alweendo said during a 2021 interview with CNN. "There is a feeling from developing countries that somehow the resources that were used to develop the Western

Hemisphere are suddenly now not the right thing to do and we need to do something else."[3]

I expect equally negative global reactions to TotalEnergies' discovery of significant light oil and associated gas off the coast of Namibia in February 2022, and Shell's January 2022 discovery of a well containing as much as 350 million barrels of oil and gas offshore from Namibia.[45]

Handwringing over African oil and gas discoveries seems to be the new norm. Nevertheless, other African countries are on the verge of experiencing similar opportunities—many with the potential to usher in valuable economic opportunities for Africans of all walks of life. Recent condensate discoveries in South Africa, for example, could represent a real turning point for the country, which has, until recently, relied on oil and gas imports and coal production for energy.

What's more, many African states have invested time and effort into developing fiscal and regulatory policies designed to encourage ongoing investments and develop well-paying jobs, capacity, business opportunities, technology sharing, and monetization. In other words, governments have been paving the way for oil and gas development to offer tangible, long-term benefits to their people.

For example, since 2020, Gabonese president Ali Bongo Ondimba has been working on economic recovery from the COVID-19 pandemic based on sustainability, transparency, and a healthy business environment. And the country's new Hydrocarbons Code, enacted in 2019, offers a more competitive fiscal regime—making Gabon's oil and gas projects more attractive while promoting the development of marginal fields by local companies.

African countries need time to seize the opportunities their oil and gas resources are delivering, despite the mounting pressure for African oil and gas activity to come to a grinding halt due to global climate change.

The narrative that African oil and gas activity is a scourge that the world must overcome is false, and frankly, insulting. African countries are free nations, not colonial territories. African leaders are more than willing to cooperate with global efforts to reach emissions standards, but they will not stand for bullying.

Africa's oil and gas discoveries can pave the way to better financial futures for millions of Africans. Many of our countries are better positioned than ever to capitalize upon these opportunities. Whether the global community likes it or not, we're going to pursue them.

We're Not Outliers

African countries are hardly alone in their refusal to accept global pressure to rush their transition from fossil fuels to renewable energy sources like solar, wind, and hydrogen power.

In May 2021, the International Energy Agency (IEA) report *Net Zero by 2050: A Roadmap for the Global Energy Sector*[6] called for a halt to oil and gas exploration around the globe at the end of the year. This dramatic measure, the IEA argued in the report, is the global energy sector's only hope of achieving net-zero emissions (ensuring the amount of greenhouse gases being emitted into the atmosphere equals the amount being removed) by 2050, a goal outlined in the Paris Climate Accords.

While some nations have put their support behind the IEA's recommendation, a number of oil- and gas-producing nations have firmly and unapologetically rejected it. "We are bringing emissions down," Angus Taylor, Australia's Minister for Energy and Emissions Reduction, said at the time the report was released. "But we're going to do it in a way that ensures we've got the affordable power that Australians need."

Taylor was not the only global leader unwilling to accept the premise that the IEA's way was the *only* way to protect the environment.

Akihisa Matsuda, the Deputy Director of International Affairs at Japan's Ministry of Economy, Trade and Industry (METI), told Reuters that his government had no plans to immediately stop oil, gas, and coal investments. "The report provides one suggestion as to how the world can reduce greenhouse gas emissions to net-zero by 2050, but it is not necessarily in line with the Japanese government's policy," Matsuda said. "Japan needs to protect its energy security including a stable supply of electricity, so we will balance this with our goal of becoming carbon neutral by 2050."[7]

Norway Oil Minister Tina Bru pushed back against the IEA's recommendations too. "It would not if Norway discontinues production," Bru said. "It would just move to other countries, and then we are no further. This is a complex global problem that requires many solutions."[8]

Similarly, in the Philippines, Energy Secretary Alfonso Cusi said a rush to cut off oil, gas, and coal financing would "set back the Philippines' aspiration to join the ranks of upper middle-income countries."[9]

I respect each of these stances and wish that Africa's oil- and gas-producing nations would be afforded the same consideration when the continent's leaders express similar viewpoints regarding forcefully shifting Africa from fossil fuels to renewable energy.

Unfortunately, the IEA is far from the only voice calling for African oil- and gas-producing countries to adopt the same timetable as developed nations around the globe. Foreign investment in African oil and gas production has been declining significantly in the last year, and when companies and financial institutions publicly consider investing in Africa's oil and gas industry, they're threatened with boycotts.

Preventing Africa from exploiting its energy resources is tantamount to leaving much of the continent not just energy-poor, but also economically impoverished. And when you factor in the fact that Africa produces only 2% to 3% of the world's CO_2 emissions from energy and industrial sources, it's clear that this kind of pressure is not only unfair but also unreasonable.

For these reasons, I continue to be excited about recent oil and gas discoveries in Africa.

Recent Discoveries

As far as I'm concerned, new oil and gas discoveries are promising developments for African countries, communities, businesses, entrepreneurs, and young people hoping to build bright futures for themselves.

Each of the recent discoveries in Africa can, both directly and indirectly, lead the way to jobs that pay well, a workforce with valuable new skills, gas for domestic gas-to-power programs, and so much more.

Here are just a few of the most promising discoveries:

Angola: In April 2021, Italian oil company Eni made a light oil discovery in Angola's deep offshore Block 15/06 with a potential totaling 200 million to 250 million barrels.

Angola, by the way, is another leading example of a country that has taken measures to make exploration and production more attractive to international oil companies. President João Lourenço has incentivized investors by offering improved contract terms to increase profitability for investors, tax relief, and the appointment of a standalone oil industry regulator. These efforts are paying off. Angola saw crude oil exports increase by 3% in the fourth quarter of 2021. The country exported roughly 394.22 million barrels of crude oil in 2021 for an annual gross revenue of USD27.87 billion, a 51.4% increase over 2020.[10]

For the first time since 2018, TotalEnergies is drilling in Angola and has executed a sale and purchase agreement with Sonangol for two blocks in Angola's offshore Kwanza Basin. Other majors, including ExxonMobil, Chevron, and BP, are also active in Angola, along with Eni.

Côte d'Ivoire: In 2021, Eni discovered the Baleine Field, which contains as many as two billion barrels of oil and nearly 2 trillion cubic feet (cf) of gas offshore of Côte d'Ivoire.[11] This is a big deal for Côte d'Ivoire, which up to now has been producing about 34,000 barrels of crude per day from four blocks.[12]

Sergio Laura, Eni's representative in Abidjan, the capital of Côte d'Ivoire, has pointed out that Eni's exploration program in-country was successful because of the cooperation of the General Directorate of Hydrocarbons, an arm of the Ministry of Petroleum, Energy and Renewable Energies, and the national company Petroci.

Going forward, Ivorians will be reaping the benefits of Eni's activities in their country. "The Italian hydrocarbon giant plans to rely on Ivorian engineers' and technicians' high level of expertise so that it can rapidly increase the responsibilities of local managers, just like it did in Ghana," Christophe Le Bec wrote for *The Africa Report*. "In Ghana (where Eni already is active), the proportion of local managers has risen from 50% to 80% in 5 years, thanks to accelerated training programmes—in Italy and elsewhere in Africa—and mentoring."[13]

Egypt: In February 2022, Dragon Oil of Dubai announced an oil discovery in the Gulf of Suez that may contain as many as 100 million barrels. Dragon Oil is a subsidiary of state-owned Emirates National Oil Co. Egyptian Petroleum Minister Tarek El Molla has said about 45 million to 50 million of the barrels can potentially be extracted, most likely in less than a year.[14]

Senegal: Another example is the relatively recent discovery in Senegal of two massive natural gas offshore fields. The projects currently being developed are predicted to come online in 2022 and 2026.[15]

The nation is taking its environmental duties seriously, too. Recently, it was announced that two floating storage and regasification units will be utilized to receive the liquified natural gas (LNG) and convert it back into a gaseous state. The offshore ships will help reduce the environmental impact that building an onshore plant would have had.[16]

South Africa: Recent discoveries of gas deposits are causing a major wave of excitement in the country. For instance, when TotalEnergies discovered gas field Brulpadda in the Outeniqua Basin in 2019, the company determined that it held at least one billion barrels of oil equivalent of gas and condensate light oil.[17]

Subsequently, in October 2020, TotalEnergies discovered gas condensate on the Luiperd Prospect, and the well was even larger than the main reservoir at Brulpadda.

All of these discoveries can help South Africa achieve its goal of building modern infrastructure while making the transition from oil to gas for powering electricity.

Major discoveries also have been made in Gabon and Ghana in 2021. In August 2021, BW Energy discovered oil in the Hibiscus North exploration well in Gabon's Dussafu Block. And in 2021, Eni announced that its Eban exploration prospect offshore from Ghana could contain 500 million to 700 million barrels of oil equivalent.[18] Because the discovery site is near existing infrastructures, Eni says, it can be fast-tracked to production,[19] which is exciting news for Ghana.

African Business Matters

In spite of the examples enumerated above, it is important to remember that not all oil and gas companies in Africa are foreign-owned. In fact, African-owned companies comprise the full range of oil and gas businesses—from upstream operations to small and mid-size service companies.

African businesspeople have worked hard to create jobs and opportunities for other African entrepreneurs as well as their fellow citizens.

For instance, the Springfield Group located in Ghana was granted the opportunity to develop the West Cape Three Points Block 2,[20] which is expected to provide many jobs for the people of Ghana.

Harlequin Oil and Gas (HOG), a Ghanaian-owned engineering services company, was recently awarded the Indigenous Company of the Year Award. The owner, Frederick Hermann Hesse-Tetteh, said that the company will "continue to employ a considerable number of permanent staff from the western region."[21] The African-owned company also sponsors development programs that train university graduates, engineers, and technicians.

Foxtrot is another African-owned business that works in upstream development on the Côte d'Ivoire. It hires many people, including scientists such as geologists, geophysicists, gas and oil architects, and reservoir engineers, along with technicians, economists, and financiers.

So, when we discuss the importance of Africa's oil and gas industry, we need to include its value to local companies, entrepreneurs, and employees at all levels.

We're Not Giving Up the Wheel

When I wrote *Billions at Play: The Future of African Energy and Doing Deals* a couple of years ago, my goal was to create a roadmap that African petroleum-producing countries can use to break away from the resource curse and fully harness the benefits of our natural oil and gas resources.

Now, suddenly, we're being told the road is closing. Immediately. Permanently.

Never mind that African countries have been putting in the hard work of building a better future for themselves. Never mind that African leaders have done their research and have started revising the fiscal and government policies to attract investors. Forget those who've revamped their governance structures to ensure that oil and gas revenue is used for infrastructure and other developments that can strengthen African economies.

Here's our response: Forget it!

We are not going to shrug our shoulders and go along with the international community's time frame for phasing out oil and gas production.

Of course, African countries will support global efforts to protect the earth from climate change. But we're going to do it our way and on our timetable.

Chapter 6

THE VITAL IMPORTANCE OF NATURAL GAS

About 11 years ago, when long-time energy sector consultant Kofi Nketsia-Tabiri began formulating an idea for a Ghana-based liquified petroleum gas (LPG) business, he decided to spend a year talking to potential customers and gaining insight into the LPG market.

What Nketsia-Tabiri learned was that consumers in rural areas assumed LPG was too expensive to meet their energy needs. Others had no way of accessing it. And in some areas, people didn't even know LPG was an option. (Not to be confused with liquified natural gas (LNG), LPG refers to propane, butane, and combinations of the two. It's produced during natural gas processing and crude oil refining and is most often used for cooking, powering heating appliances, and fueling vehicles.)[12]

After receiving this feedback, Nketsia-Tabiri resolved to build a company that would make LPG affordable and easily accessible to Ghanaians wherever they lived, from urban areas to remote locations. By doing so, he hoped to enable Ghanaian households to stop resorting to such commonly used—but hazardous—energy sources as charcoal, kerosene, and wood for cooking and heating their homes.

XpressGas Limited, the company Nketsia-Tabiri founded, is now a thriving LPG supplier, distributor, and marketer that serves Ghanaian households, institutions, and the commercial sector. It employs more than 270 people and has refilling stations in all 10 regions of Ghana.

XpressGas is known for its outside-of-the-box business models. Its innovative Swap N Go program, for example, offers paid subscriptions that include unlimited technical support and free refill deliveries. In an article about Nketsia-Tabiri for an African business publication *How We Made It In Africa*, Jeanette Clark wrote that "the company goes to the end user's address to swap the empty canisters for full ones after payment. The canisters remain the property of XpressGas, taking away one of the barriers to entry for consumers who cannot afford to buy their own."

At the 2019 Ghana Oil and Gas Awards, Swap N Go was named Innovative Product of the Year, and Xpress Gas Limited was named Emerging Brand of the Year. Since then, Xpress Gas Limited has begun testing another program developed with customers' needs in mind: a pay-as-you cook option. Customers prepay for gas on their mobile phones, which allows them to activate a valve on their LPG canister to provide the purchased amount.[3]

"This is a low-margin, high-volume business that targets end-users on the lower level of the energy pyramid who would typically use charcoal for cooking, but who can afford only small amounts of charcoal at a time," Nketsia-Tabiri said. "They would not have the capital to buy full cylinders of gas. In Ghana, we have close to two million households that don't have access to clean cooking fuels. In Africa, most of the sub-Saharan African countries still depend on traditional fuels. We will initially focus on Ghana and then, the world is ours."

I love Nketsia-Tabiri's confidence—and his devotion to his fellow Ghanaians. Not only is XpressGas a strong example of the potential of the free market to empower African people, communities, and businesses, but it also illustrates the valuable role natural gas and gas products can and should play in fueling those efforts. Gas operations contribute to economic growth by creating jobs and generating tax revenue that can be used to build much-needed infrastructure, train Africans for energy-industry careers, and pursue technological advancements. Natural gas as a feedstock for LPG, LNG, and petrochemicals supports further growth and economic diversification. And natural gas-powered electricity plants will play a vital role in alleviating Africa's widespread energy poverty crisis: the millions who lack access to reliable electricity.

As I have already shared, I am strongly in favor of capitalizing on Africa's approximately 600 trillion cubic feet of proven natural gas reserves.[4] Despite the pressure to rush to renewable energy sources, the promise of natural gas has not evaporated. Far from it. Our natural gas resources will be vitally important to meeting the energy demands of our continent's rapidly growing population. By 2050, Nigeria alone is predicted to be the third most populous country in the world. Three cities in Africa—Cairo, Egypt; Lagos, Nigeria; and Kinshasa, Democratic Republic of Congo—each already have 10 million or more residents. And more cities, including Nairobi, Kenya; Dar es Salaam, Tanzania; Khartoum, Sudan; and Casablanca, Morocco, will soon be able to say the same.

Natural gas is efficient—just a small volume generates a great deal of electricity—and when burned, it releases less carbon dioxide, less sulfur dioxide, and less particulate matter per unit of energy than liquid hydrocarbons or coal. Natural gas also can be integrated with renewable energy sources, like solar power and wind, to allow for minimal carbon emissions.

I want to be clear: I do realize that natural gas production, processing, transport, and usage release methane, a potent greenhouse gas, into the atmosphere. But I am confident that harnessing natural gas in concert with methane mitigation measures—beginning with monitoring and repairing leaks—will allow us to minimize our impact on the climate while growing our economy and meeting our people's energy needs.

These are just a few of the reasons I'm convinced that natural gas must be a major component of an energy mix that paves the way for a safe, stable, and prosperous future in Africa.

Here's a closer look at some of the ways gas can position African countries for a successful energy transition.

Vital Gas-to-Power Programs

With energy poverty continuing to leave roughly 600 million people in Sub-Saharan Africa in the dark, we absolutely must continue to pursue each and every solution available to us. Could that include harnessing wind power? Absolutely. Bring it on. Microgrids? Let's do it. But at the same time, using our abundant natural gas resources to power electricity generation remains one of our best options in Africa.

Not only does natural gas produce less carbon dioxide than other fossil fuels, but gas-fired electricity facilities also can be built more quickly, and at a lower cost, than coal or nuclear plants. In addition, natural gas can overcome the shortcomings of renewable energy sources like wind and solar power by being used *with* them to generate electricity. As Chris Mooney wrote for *The Washington Post*, "Because of the particular nature of clean energy sources like solar and wind, you can't simply add them to the grid in large volumes and think that's the end of the story. Rather, because these sources of electricity generation are 'intermittent'—solar fluctuates with weather and the daily cycle, wind fluctuates with the wind—there has to be some means of continuing to provide electricity even when they go dark. And the more renewables you have, the bigger this problem can be."[5]

More than a dozen African states are successfully utilizing natural gas that they produce themselves or import from other African countries to generate electricity. Look at the success of gas-to-power in Côte d'Ivoire, a country that has become one of the top electricity producers in West Africa over the last decade; it even sells surplus electricity to six other African countries. Much of the country's electricity is generated by gas-powered plants. Energy access in Côte d'Ivoire has been increasing since 2012 when the government introduced regulations that encouraged private infrastructure investments. Thanks to those efforts and public-private investments, access to electricity in Côte d'Ivoire surged from 34% of the population in 2013 to approximately 94% today.[6]

Natural gas also is part of South Africa's strategy to reduce carbon emissions to net-zero by the year 2050. Yes, South Africa plans to build more wind- and solar-powered energy plants, but it also will rely on gas-powered facilities to achieve its objective.

"Gas is seen as a bridge to transitioning South Africa away from coal power while allowing the country to maintain its goals of economic generation through industrialization," Shridaran Pillay, Africa Director at risk advisory service Eurasia Group, told *Bloomberg*. "I do not think that renewables in South Africa could be scaled to a level where they would provide the same opportunity to replace base-load generation in a way that gas can at this stage."[7]

With major natural gas discoveries in Mozambique, Ghana, Senegal, and Tanzania, we can—and should—expect to see even more African countries harnessing their natural resources to provide their populations with reliable electricity. Projects on the horizon include Nigeria's Kingline Power Project, a 550MW gas-to-power facility; the Temane Thermal Power Station, a 450MW gas-fired power plant in Mozambique; and the Ghana 1000 Power Station, a 1,300MW gas-fired power plant in the western portion of the country.

Gas-to-power projects benefit countries and communities, but they are also meaningful in the lives of average people. Providing people with electricity plays a role in alleviating poverty and reducing unemployment. Energy opens the door to opportunity.

Ensuring widespread access to electricity must be our priority in Africa. It will be impossible to have a just energy transition if we fail to do everything possible to address energy poverty. And "everything" includes harnessing natural gas.

But natural gas is more than the solution to our energy struggles in Africa. It also offers a pathway to prosperity.

Creating New Revenue Streams

You may remember that I already mentioned once or twice the vital importance of monetizing our natural gas. Why do I emphasize that point? Because monetization—deriving value from every step of a natural resource's value chain, from extraction to the use of a final product—is one of the best things we can do today to help ourselves tomorrow. Monetizing our natural gas in Africa paves the way for economic growth and the establishment of new companies in multiple industries. Those companies will need permanent employees and leaders, along with supplies and services.

Monetization will also help generate revenue to channel into the infrastructure new companies need to be competitive, like roads, pipelines, and ports, resulting in even more economic growth and diversification. All of this equates to greater opportunity for African people and businesses.

Monetization, by the way, is not an abstract theory. In my last book, I wrote about Equatorial Guinea's initial efforts to cultivate a world-class natural gas sector. The country's Minister of Mines and Hydrocarbons, Gabriel Mbaga Obiang Lima, has been working since 2018 to develop a "gas mega hub," an offshore and onshore network of production, processing, and transport facilities that will monetize both domestic offshore gas reserves and neighboring reserves.[8]

Those efforts began almost immediately with the USD330 million Alen Gas Monetization project. When Equatorial Guinea signed an agreement with Texas-based Noble Energy (now owned by U.S. multinational Chevron), it created a framework for the development of natural gas from the offshore Alen Field and outlined commercial terms for Noble to provide 600 billion cubic feet (bcf) for onshore LNG processing facilities.[9]

Now, those efforts are making important strides forward. In early 2021, Chevron achieved first gas flow from the project. A 70-kilometer, 950-million-cubic-foot pipeline has begun transporting natural gas from Alen Field to Equatorial Guinea's LNG production facility at Punta Europa,[10] as well as the Alba LPG plant. Going forward, revenue from these operations can be used to build pipelines, platforms, ports, and other infrastructure to support Equatorial Guinea's gas sector, draw more foreign investment, and foster more economic growth and diversification.[11]

There's also a lot of industry buzz about monetization potential in South Africa since TotalEnergies made significant gas condensate (light, liquid hydrocarbon) discoveries there in 2019 and 2020. The Brulpadda and Luiperd Prospects are located on Block 11B/12B in the Outeniqua Basin, 175 kilometers off the southern coast of South Africa. It's estimated that the two fields could hold up to one billion barrels of gas condensate each, with the potential to move the country away from heavy coal usage and gas imports well into the 2020s.[12] The potential of this discovery is huge.

These prospects could lead to large local investments that could help South Africa develop the infrastructure it needs to develop a successful natural gas sector and shift away from its dependence on coal, which currently powers 87% of South Africa's power grid. The gas produced could support gas-to-power projects and be used to make gas products, supporting economic growth and industrialization.

It is now up to government leaders to move quickly so South Africa can capitalize on these discoveries. Leaders need to ensure that labor laws are fair for everyone and don't impede foreign investment. They should consider tax incentives for skills training so that South Africans reap the benefits a thriving gas industry can offer while minimizing red tape for international oil companies. The investment required to extract this gas likely would range from USD40 billion to USD60 billion, of which USD35 billion would be spent locally. South Africa could be a shining example of monetization.[13]

We have other examples of successful gas monetization in Africa to look to as well:

- The Nigerian Independent Petroleum Company (NIPCO Plc.) has been developing compressed natural gas (CNG) infrastructure and dispensing outlets that allow the public to fuel their cars with CNG instead of gasoline. Since this project began in 2009, NIPCO has opened CNG stations in Benin, Edo State, and Ibafo and established a gas compression plant in Ogun State.[14,15] NIPCO's success with these projects also allowed it to construct a 10,500 metric-ton LPG storage facility.

- Through its subsidiary Gaz Du Cameroon, Victoria Oil & Gas has built a fully integrated gas network in the port city of Douala, Cameroon. The gas company supplies domestic natural gas for about 20 companies in Douala, supporting industrial and manufacturing activities there.[1617]

- Additional efforts in Equatorial Guinea include the construction of a regasification terminal to process natural gas across several industries, including cement, on the mainland. In Abidjan, the government has been pushing for the procurement and deployment of CNG buses that also run on domestic natural gas.

These projects are only a drop in the bucket in terms of what natural gas can do for us. I also see great potential for it to drive economic diversification as a feedstock for new, value-added products, including petrochemicals.

Petrochemicals: A Formula for African Success

In late 2020, the West African Students' Union (WASU) president described Nigerian business magnate Aliko Dangote as a godsend. That's because his industrial conglomerate, Dangote Group, was providing something young West African adults desperately need: jobs. The company was employing more than 200,000 West African youths at the time.[18]

"As students, we are mostly scared of our future because of the high unemployment rate in our region, but with a man like Aliko Dangote, we are optimistic," WASU President Comrade Romans Pillah said shortly after the student union named Dangote Group the "Most Outstanding Company in Africa" for job creation.

Dangote Group comprises 18 subsidiaries across a wide range of sectors, and it serves as a strong example of what economic diversification can do for Africa. I'm especially excited about Dangote Refinery and Petrochemicals, which has the potential to create thousands of additional jobs—estimates range from more than 30,000 jobs to 300,000[19]—when completed in 2022.

The petrochemical industry represents economic opportunity around the globe, including in Africa. A recent report by professional services network Deloitte noted that even when demand for automotive fuel declines during the next decade as electric vehicle usage increases, petrochemicals will be needed. The huge array of products made with petrochemicals is too long to list in this book, but it ranges from food preservatives and shampoos to detergents, lubricants, and fertilizers.[20]

"Petrochemicals are expected to be the largest driver of world oil demand growth, surpassing that of gasoline or diesel by 2030," the Deloitte report said. "Petrochemicals will likely remain the mainstay of all products, industrial or consumer, which the world consumes on a daily basis. It then seems natural for major refining and integrated oil and gas players to reconfigure their plants towards producing more petrochemicals rather than just fuels."[21]

Using natural gas as a feedstock to produce fertilizers and fertilizer ingredients, including urea, ammonia, and NPK (nitrogen, phosphorus, and potassium) would be a particularly good fit for African nations, considering the fact that agriculture is so important in our continent. In Sub-Saharan Africa alone, small farmers comprised more than 60% of the population in 2019, and about 23% of the region's GDP came from agriculture.[22]

In the March 2020 issue of *World Fertilizer*, Contributing Editor Gordon Cope wrote that opportunities for the fertilizer sector abound in Africa. In 2019, our continent's population exceeded 1.32 billion people. At its current growth rate of 2.5%, the population is expected to almost double to 2.5 billion by 2050. Cope estimated that NPK consumption in Africa was low compared to overall global usage. "However, due to initiatives in many different countries, the creation of fertilizers suited to African soils (as well as technical expertise and availability of farm credit) is expected to rise significantly," Cope wrote.[23]

When it comes to fertilizer and other petrochemical plants, we can't rely upon an "if we build it, they (buyers) will come" strategy. Unless we build and run our petrochemical industry in a cost-effective manner, our products will be too expensive to compete in the global market. We already have the advantage of ample natural gas reserves, but we must consider all the pieces of the puzzle, including developing supply chains and the necessary infrastructure for petrochemical plants to efficiently manufacture and distribute their products. We also need the necessary technical expertise and skilled employees.

So how do we get there? For starters, gas-producing countries must make it a priority to ease up on gas exports and gas flaring and, instead, commit to dedicating their gas resources for domestic needs. We need a strong regulatory framework to foster the creation of a petrochemical industry. Solid, business-friendly regulations will let both domestic and foreign companies know what to expect, from hiring requirements to environmental regulations, if they want to develop a petrochemical plant.

I'd like to see governments go a step further by offering tax packages that incentivize investments in fertilizer plants and infrastructure. In some cases, an African state might have to consider providing backing, such as a guarantee, to ensure that the necessary financing is approved. And, finally, we need the support of the private sector and public-private partnerships to cultivate the technical knowledge and skills we need. As plants begin generating revenue, a portion of it can be invested to develop even more infrastructure and support local capacity-building programs.

I am seeing signs that we're moving in that direction. In fact, as I've been writing this book, the Dangote Fertiliser Plant in Nigeria (yet another Dangote project) began production and rolled out its first bags of fertilizer. The USD25 billion granulated urea fertilizer plant was developed in Ibeju Lekki, Lagos State by Dangote Industries Limited.

Also exciting are Brass Fertiliser and Petrochemical's (BFPCL) plans to build a natural-gas-to-methanol plant capable of producing 10,000 tons per day in Odioma, Brass Island, in Nigeria's Bayelsa State. In an agreement with the Nigerian government, BFPCL has offered a 1% stake in the plant to its host communities. Plans for the USD3.5 billion project estimate that it will create 30,000 jobs (direct and indirect) during construction, and that the plant will employ 5,000 people when completed.[24]

"The methanol plant was in line with Nigerian President Muhammadu Buhari's plan to aggregate and monetise all stranded gas in the Brass area, which amounts to over 10 trillion cubic feet of gas, into the processing facilities to be built in the hub," Nigerian Minister of State for Petroleum Resources Timipre Sylva said in February 2021.[25]

Not only could an increase in fertilizer usage in Africa produce economic benefits, from job creation to greater agricultural crop yields, but it also could have environmental benefits. Raising crop yields helps prevent cropland expansion into forests, which play an important role in preventing climate change by absorbing fossil fuel emissions.

Promising Liquid Assets: Liquified Natural Gas

A growing petrochemicals/fertilizer industry is only one of the many opportunities natural gas puts within African countries' reach. Let's take a closer look at LNG. Technological advancements have made processing it (converting natural gas to a liquid state), transporting it, and degasifying it more affordable than ever. That represents tremendous promise for our continent. LNG for gas-to-power projects can be imported from gas-producing African states to those that lack it. Gas-importing countries also can reap economic benefits from the construction of regasification facilities, along with LNG storage terminals and transportation infrastructure.

What's more, the opportunities associated with LNG aren't limited to huge companies. Increasing numbers of African entrepreneurs are embracing small-scale LNG (SSLNG), the transport and supply of small to medium quantities of LNG.[26]

Gas Processing News recently described SSLNG as a potential game-changer for the global energy industry. Demand for LNG is increasing because it's cheap and clean. It emits approximately 20-30% less greenhouse gas than diesel or gasoline as a transportation fuel and 50% less greenhouse gas than coal to generate electricity. While large-scale

LNG plants are being developed around the globe, these projects are complex and take years to complete. SSLNG projects, on the other hand, can be launched more quickly and are more manageable for smaller companies.

"The ability to produce LNG at remote locations, and the availability of technologies to conveniently transport the product, has made small-scale LNG an attractive way of delivering natural gas to areas of demand," wrote *Gas Processing News*. "The USD25 billion SSLNG market is forecast to double over the next decade."

An additional plus: Small-scale operations can transport LNG in its liquid form and leave regasification to their customers, reducing SSLNG operators' capital expenditures.[27]

We're already seeing successful SSLNG operations in Africa. As of summer 2020, Golar's Hilli Episeyo, the world's first converted floating LNG (FLNG) vessel, had provided more than 2.5 million tons of fuel since beginning commercial operations 2 years earlier.[28] In July 2021, Golar LNG reached an agreement with Perenco and Cameroon's national oil company SNH to increase utilization of Hilli Episeyo. In 2022, the capacity utilization of Hilli Episeyo will increase by 200,000 tons of LNG, bringing total utilization to 1.4 million tons.[29]

SSLNG represents a valuable opportunity for small African companies that cannot raise large amounts of capital or purchase big-ticket items. They can run LNG programs on a small scale with barges and deliver LNG across the continent to provide power and meet growing domestic demand.

This is not to say that large-scale LNG projects don't have value for Africa. These projects, from the Greater Tortue Ahmeyim LNG project off the shores of Mauritania and Senegal to the Tanzania LNG Liquefaction Project, will create permanent well-paying jobs. They also represent opportunities for African suppliers and service providers. And the revenue they generate can be used to develop infrastructure and launch capacity-building projects, which will contribute to more economic growth.

LNG offers these benefits and opportunities with minimal impact on the environment. LNG releases 45-50% less carbon dioxide than coal and 30% less carbon dioxide than fuel oil. It reduces nitrogen oxide emissions and produces minimal particulates compared to other fuels.[30]

Yes, natural gas production and processing release methane, but the impact of methane emissions on the environment can be minimized with current and developing solutions.

Emissions Control

In 2021, discussions about climate change increasingly centered on the role methane plays in warming our planet. This greenhouse gas is more powerful than carbon dioxide at trapping heat (how much more is a little fuzzy; I've seen numbers ranging from 36 times[31] more power to 86 times[32]), but it doesn't carry the same long-range threat as coal or oil. Much of the methane released today will no longer be in the atmosphere a couple of decades from now, while carbon dioxide remains for centuries.[33,34]

That's not to say that minimizing methane emissions isn't important. A 2021 United Nations climate report has called for reducing methane emissions 40-45% by 2030 to limit the global average temperature rise to 1.5 degrees C. Not all of the responsibility will fall on the oil and gas sector, which is responsible for 35% of the world's human-made methane emissions. Agriculture is responsible for 40% and organic waste for 20% of methane emissions.[35]

The good news for the gas sector is that minimizing methane emissions is doable. There are effective steps that can be (and are being) taken now, including the detection and repair of methane leaks at gas facilities and infrastructure. The UN Environment Programme

(UNEP) has estimated that 60% of targeted measures like these are low cost, and 50% have negative costs, meaning companies can increase their revenues by preventing leaks and capturing methane instead of releasing it into the atmosphere.[36]

And increasing numbers of companies are committing to step up their efforts in this area. In late 2020, national oil companies (NOCs), international oil companies (IOCs), and majors around the globe joined the Oil and Gas Methane Partnership, which committed to reducing methane emissions by 45% by 2025 and by 60% by 2030 and to adopt more stringent reporting measures. In doing so, these companies can reduce methane emissions from 13% of total upstream emissions to less than 5% by 2030, according to global energy consultant Wood Mackenzie.

"We expect more oil and gas producers to set their own methane reduction targets, expand methane reporting in their sustainability reports, and roll out methane monitoring and reduction technologies," a recent Wood Mackenzie report stated. "We also expect routine flaring to decrease across the world, particularly in the U.S."[37]

Flaring is the controlled combustion of natural gas, and venting is the deliberate release of natural gas into the atmosphere. Both practices occur regularly during oil and natural gas drilling, production, gathering, processing, and transportation. Not only are these practices environmentally harmful, but they also waste valuable natural resources. A number of African leaders recently agreed to endorse the "Zero Routine Flaring by 2030" initiative introduced by the World Bank. They include Angola, Egypt, Cameroon, Gabon, Morocco, Niger, Nigeria, Republic of Congo, and South Sudan.[38] I expect even more African oil- and gas-producing nations to commit to reducing or altogether ending their flaring and venting so they can continue growing and strengthening their gas sectors.

Meanwhile, we're seeing increasingly innovative—and frankly, cool—solutions being developed to enhance leak detection programs.

Colorado-based LongPath Technologies, for example, has been testing a frequency comb laser for this purpose. *GreenBiz* recently featured LongPath's chief technology officer, Greg Rieker.

"Rieker is betting that his laser-based system can catch leaks more effectively and affordably than the current industry standard, which involves using a $100,000 infrared camera to randomly scan a well pad's welter of valves, tanks, separators and compressors a few times a year," Jonathan Mingle wrote for *GreenBiz*. "The camera's ability to detect leaks is highly dependent on weather conditions and the operator's skill; it also only collects data as a single snapshot in time and doesn't indicate the size of a leak. Rieker's frequency comb laser, meanwhile, sweeps continuously across the plain for weeks and months, hunting for and tallying up errant methane molecules."[39]

If laser methane emission detection isn't techie enough for you, how about catching leaks from space? In recent years space agencies and companies around the globe have launched satellites to map and measure methane emissions. And more projects are on the way, including Carbon Mapper, a public-private partnership based in California, and MethaneSAT, a project of America's Environmental Defense Fund.

"MethaneSAT will focus on the global oil and gas industry and aims to be sensitive enough to reveal the multitude of small methane releases that can account for the majority of emissions," reported Cheryl Katz for PBS. "The findings will be made available to industry operators, regulators, investors and the public in near-real-time. The data," she said, will help "prioritize what makes the most sense in terms of emissions reductions and mitigation."[40]

Drone-based systems are taking off, too. SeekOps, for instance, said the miniaturized methane sensor it has developed is sensitive enough to detect gas in the parts-per-billion range and compact enough to be mounted on a lightweight drone. Shell Oil has been harnessing the technology of Avitas, a Baker Hughes company, to conduct drone-based methane leak and repair operations in the U.S.[41]

Not only do we have options available for limiting methane releases, but we're also seeing advancements in technology to limit carbon emissions.

How about using carbon capture and storage (CCS) technology, which could minimize the carbon footprint of gas-fired electricity plants? This technology captures CO_2 at the source, transports it to storage locations, and sequesters it there. CCS is an expensive technology, but it could represent an excellent opportunity for investments, especially with interest in and demand for carbon capture surging. What's more, storing CO_2 below ground requires compatible geology, and parts of Africa could potentially meet that need.

My point is this: Africa can be part of the solution when it comes to cultivating environmentally friendly natural gas operations.

I'm excited about what natural gas is starting to accomplish in Africa. I'm excited about the role it will play in eradicating energy poverty and generating the revenue we need to build up our infrastructure, grow our economies, and create jobs and business opportunities for our people. We will realize the economic, environmental, and social benefits of gas production and monetization. We're going to take advantage of gas products to meet pressing needs, from safe cooking fuel to agricultural fertilizer. And by doing all this, we're going to enable African countries to embrace a future of renewable energy sources from a place of stability and prosperity.

Chapter 7

HYDROGEN CAN POWER TREMENDOUS GROWTH FOR AFRICA—IF WE'RE SMART ABOUT IT

IN THE NOT-SO-DISTANT future, South Africa could be home to Africa's first "hydrogen valley," an incubator of sorts for pilot projects across the hydrogen value chain: production, transport, distribution, and end-use. Hydrogen valleys already exist in Asia, Europe, and North America, among other places. Once completed, South Africa's hydrogen valley project could be a major gateway to economic growth and job creation in the region.

South Africa and its partners—mining company Anglo American Platinum, local fuel cell company Bambili Energy, and French energy group Engie SA—completed their feasibility study for the hydrogen valley in October 2021. The study identified a wide range of promising pilot projects, including converting buses, mining trucks, and forklifts into hydrogen-powered vehicles; developing ethylene and ammonia from green hydrogen (hydrogen produced with renewable energy sources); using green hydrogen to power steel production;

building green hydrogen-powered paper mills; and powering corporate office buildings with hydrogen fuel cells, among others.[1]

The hydrogen valley project is partially the result of the South African government's strategic approach to a just energy transition that fosters economic growth. In addition, the South African Cabinet has approved a National Hydrogen Society Roadmap, and Phase 3 of the country's Economic and Recovery Plan includes the creation of a South African hydrogen industry.

"The implementation of Phase 3 of the Economic Reconstruction and Recovery Plan is driven by the core elements of 'reconstruct' and 'transform,' and this entails building a sustainable, resilient and inclusive economy," the Director-General of South Africa's Department of Science and Innovation, Dr. Phil Mjwara, told business news website BusinessTech. "The establishment of a South African hydrogen valley is therefore seen as an opportunity that has great potential to unlock growth, revitalize the industrial sector, and position South Africa to be an exporter of cost-effective green hydrogen to the world."[2]

This is just one example of what hydrogen power projects can mean for our continent—and how governments can take a proactive, planned approach to successfully developing a hydrogen-based economy. In South Africa's case, the country's hydrogen aspirations are linked to the country's massive reserves of platinum, a key component in hydrogen fuel cells. But platinum is hardly a requirement for successful African ventures.

The opportunities posed by hydrogen are too huge to ignore. Countries around the world look to this clean energy source to meet their net-zero emissions goals. The International Energy Agency (IEA) has estimated that the global market for low-carbon hydrogen could expand from 450,000 tons in 2020 to 18 million tons in 2030. Demand will grow even more, according to the IEA, if the world reaches net-zero emissions by 2050, since there would be far less (if any) usage of fossil fuels.[3]

To understand why hydrogen represents such a great opportunity for Africa, we need to take a closer look at one of the most common ways it's produced. Hydrogen rarely exists on its own; it's almost always connected to other elements. In water, hydrogen is the "H" in H_2O. Through a process called electrolysis—passing an electric current through water—producers can break the chemical bond between hydrogen and oxygen to create two separate gases. Hydrogen is considered "green" when renewable energy sources power the electrolysis process. Africa is ideal for green hydrogen production because we have a wealth of renewable energy sources: ample sunlight for solar-powered electrolysis and considerable landmass, both of which are necessary for generating wind and solar energy. Africa also has the highest untapped hydropower potential worldwide, with only 11% of its potential being utilized.[4]

Africa's capacity as a green hydrogen producer is generating tremendous interest in the continent and around the world. I'm excited about hydrogen's potential, too, but with a caveat: We must ensure that African countries, communities, businesses, and individuals realize the full benefits of Africa's hydrogen activity. In spite of surging interest in zero-carbon green hydrogen, we should not turn our backs on low carbon "blue hydrogen" produced with African natural gas and carbon capture and storage (more on this later). We must develop and implement policies that attract foreign hydrogen investments without sacrificing local content. Finally, we must commit to building a domestic hydrogen economy to meet local needs and fostering sustainable economic growth, rather than focusing solely on exports.

If we do those things, hydrogen truly will help power our future and pave the way to a just energy transition.

European Excitement

Why is hydrogen considered such a bright star in the world of renewables? Like wind and solar energy, it doesn't release carbon into the atmosphere: When burned, it only emits water. Hydrogen is also a great fit for sectors that are not easily compatible with other clean energy solutions, which makes it particularly appealing to countries aiming to decarbonize. Hydrogen can power the high-temperature processes necessary to manufacture cement, steel, iron, and fertilizer. Hydrogen-based fuels can be used for shipping, aviation, and heavy trucking. Hydrogen also can generate energy for power grids and address the intermittency issues that come with relying on solar and wind for power.[5]

Hydrogen is already heavily used around the globe for industrial purposes, but most of this hydrogen is produced by natural gas or coal, meaning that greenhouse gasses are still associated with its production.

Transitioning to green hydrogen, many believe, will be the game-changer.

According to Energymonitor.ai, the European Union, along with Australia, Japan, New Zealand, and South Korea, have already launched green hydrogen strategies or made major announcements about the role hydrogen will play in their energy transitions going forward. Russia's invasion of Ukraine in February 2022 propelled European nations to lessen their dependency on Russian energy, leading interest in green hydrogen to surge to even greater levels. In a recently updated energy transition outlook, Consultant DNV projected that green hydrogen demand in 2030 will be 25% higher than in its pre-war outlook, according to Recharge.[6]

Even as countries pursue greater green hydrogen use, many are faced with the challenge of getting the volumes they'll need. Look at Germany: its government estimated in 2022 that the country will have to import 40-60% of its green hydrogen. The country's USD10.6 billion (€9 billion) national hydrogen strategy includes €2 billion in support of green hydrogen projects in partner countries around the world.[7] This is good news for Africa, since Germany is extremely interested in counting African states among its partner countries. For several years now, Germany has been laying the groundwork for green hydrogen production partnerships and projects in our continent.

One of Germany's largest initiatives in Africa is the hydrogen partnership its Federal Ministry of Education and Research (abbreviated BMBF) launched with countries in West Africa. Through that partnership, BMBF developed the Atlas of Green Hydrogen Generation Potentials in Africa, a resource that identifies suitable places for green hydrogen production in West Africa. According to the atlas, the region could generate up to 165,000 terawatt-hours of green energy a year, which is 1,500 times what Germany expects its hydrogen demand to be in 2030.

German Federal Research Minister Anja Karliczek noted that she isn't interested in simply taking what Germany needs from West Africa; she is intent on cultivating a partnership where all parties benefit equally.

"Green hydrogen offers a real chance to launch a development in Africa which is driven by the African countries themselves," Karliczek said. "The region can become the world's energy powerhouse—thanks to green hydrogen. What I would like to emphasize is that we want to import energy from Africa only when local demand is covered. A partnership between Africa and Germany is therefore a win-win situation: It means Africa can supply itself with energy while also profiting from the export of hydrogen. Meanwhile, Germany covers its own demand for green hydrogen while also profiting economically from technology exports."[8]

BMBF appears to be honoring its commitment to creating a win-win situation. For example, German research center Forschungszentrum Jülich and RWTH Aachen University, the largest technical university in Germany, designed a new two-year International Master's Program in Energy and Green Hydrogen (IMP-EG), which is designed to prepare students in West African countries for a future in green hydrogen. BMBF is funding it with €8 million initially until 2025. Plans call for 60 students from 15 African countries to participate in online training in international standards and best practices related to green hydrogen. The program also includes a semester in Germany during which students will gain practical experience and write their thesis.[9]

In addition to their partnerships in West Africa, Germany has committed USD45 million to the H2Global Development Initiative, intended to aid South Africa in developing a green hydrogen economy. Based on support from the initiative, plans are already underway for South Africa's energy and chemical company SASOL to start producing sustainable aviation fuel created with green hydrogen.[10]

In 2021, Karliczek and Obeth Kandjoze, head of Namibia's Planning Commission, signed a joint communiqué of intent to build up Namibia's solar and wind energy potential and then move on to develop Namibia's green hydrogen economy. Germany's Science Ministry will provide up to €40 million for these efforts.[11]

The Promise of Grand Inga

If everything comes together as proposed, the Democratic Republic of Congo (DRC) will become the site of the world's largest hydroelectric power plant. The project, known as Inga 3 or the Grand Inga, would be the DRC's third hydroelectric power plant on the Congo River, where it would generate a massive 44GW of electricity. A portion of the electricity generated by the Grand Inga could be transformed into green hydrogen, which could then be transported by Europe by tanker vessels. That green hydrogen would be affordable, too, produced at an estimated cost of USD 0.03/kWh.[12]

That kind of capacity has not gone unnoticed by European countries. According to the Africa-EU Energy

Partnership (AEEP), green hydrogen investments from Europe to Africa could reach €75.6 billion by 2030.

Günter Nooke, German Chancellor Angela Merkel's representative in Africa, has encouraged German companies to invest in the dam project. "Inga 3 could be the flagship project between Africa and the EU and the starting point for the industrialisation of Central Africa," Nooke said in an interview with Upstream.[13]

Phillipe Benoit, Managing Director, Energy and Sustainability at Global Infrastructure Advisory Services 2050, has expressed optimism about the project as well and has written about the importance of moving forward in an ethical way. Benoit has emphasized that project participants need to manage the project's social impacts, including the potential displacement of tens of thousands of people, along with environmental impacts such as biodiversity loss and groundwater pollution. Above all, Benoit wrote, it will be important to make sure the Congolese people don't miss out on the benefits of the Grand Inga project, including decreased energy poverty—only about 9% of the DNC's population has access to electricity—job creation, and economic growth.[14] I strongly agree.

Germany isn't the only European country interested in Africa's green hydrogen potential. Namibia is developing a hydrogen plant near the Tsau//Khaeb National Park in collaboration with Hyphen Hydrogen Energy, a joint venture between British Virgin Islands-registered investment holding company Nicholas Holdings and Enertrag, a German renewable energy provider. After the plant's first phase is completed in 2026, it will be capable of producing two gigawatts of renewable power.[15]

During the UN Climate Change Conference in Glasgow, both The Netherlands and Belgium agreed to cooperate with Namibia on green hydrogen production.[16] In addition, the Namibian Ports Authority (Namport) has signed a memorandum of understanding with The Netherlands's Port of Rotterdam that, among other things, calls for joint efforts to position the ports to become green hydrogen hubs and facilitate the flow of green hydrogen from Namibia to the Netherlands.[17]

European countries' interest in partnering with Africa is encouraging. But Africa's vast hydrogen potential isn't limited to the green variety.

Blue Goes Well With Green

In the spectrum of colors used to describe the way we make hydrogen, green refers to hydrogen produced without carbon while blue hydrogen is produced with minimal carbon. The process is powered with natural gas and utilizes carbon capture and storage technology. It only makes sense to harness Africa's abundant natural gas reserves for blue hydrogen production as well.

We need to admit that Africa's hydrogen industry likely will include both green and blue hydrogen production for quite some time. Despite the long-term potential of green energy production in Africa, relying solely on renewables to produce hydrogen is simply not practical—at least for now. Even in the U.S., where renewable generation is moving at a considerably faster pace than it is in African countries, electricity generation still requires fossil fuels and probably will continue to do so for several decades. Generating sufficient renewable energy to serve current electricity needs, and produce adequate hydrogen for domestic use and imports, likely will take considerably longer for Africa, even with the investments and support of Western countries.

Rather than seeing this as a problem, we should view it as an opportunity. As natural gas is employed for blue hydrogen production, a portion of the resulting revenue can pay for building green hydrogen infrastructure, from electrolysis and fuel cell plants to hydrogen pipelines and storage facilities. Gas revenue also can support African states in diversifying their economies with the development of green industrial operations—and Africa can begin the critical step of developing local demand for hydrogen products.

In my last book, *Billions at Play: The Future of African Energy and Doing Deals*, I pointed out the need for African states to stop exporting so much of their natural gas. It would be far more beneficial to channel it into gas-to-power programs to generate electricity for the domestic market, in addition to taking advantage of gas as inputs for value-added operations such as fertilizer and petrochemical manufacturing. The same logic applies to the green and blue hydrogen that African countries will be producing: If we focus solely on hydrogen exports, we'll be working against our own best interests.

The way I see it, Africa's 1.2 billion people are going to need more energy than anything in the decades to come; local energy consumption is going to be bigger than that of China. So, **we have to** think differently about what we do with the energy we produce, including hydrogen. Africa's hydrogen industry should not be limited to sending hydrogen to Europe from Africa and then importing finished hydrogen products back to Africa. We've done that with our petroleum products; it's a faulty strategy.

This time, let's get it right.

We also should look at providing products and services to support the hydrogen industry. And, like the hydrogen valley coming to South Africa, we should be considering hydrogen's entire value chain.

Consider fuel-cell electric vehicles (FCEVs), which are powered by hydrogen. We should be preparing now to capitalize on these vehicles as they become more affordable and demand increases. Unlike electric vehicles (EVs), FCEVs can be fueled in less than four minutes and drive about 300 miles between refuelings.[18] According to the World

Economic Forum, only 550 hydrogen refueling stations exist today. Why not build these stations in Africa?

What's more, when we think about vehicles, why limit ourselves to cars and trucks? What about FCEV forklifts? A recent report by Bain & Company titled *Business Opportunities in Low-Carbon Hydrogen* makes the case for thinking beyond personal vehicles: "Because the refueling time is much faster than for a battery, and because a fuel cell's output doesn't wane at low-charge levels, forklifts powered with hydrogen fuel cells already present a competitive option with superior performance and flexibility," the report noted. "An electrolyzer running from grid-sourced renewable electricity can produce enough green hydrogen for a fleet of forklifts. Vehicles used in mining are another example where hydrogen could make sense as a tool for decarbonizing, given the similar uptime requirements."[19]

Given Africa's widespread mining industry, and the likelihood of even more mining projects on the horizon to provide cobalt and rare minerals for EV batteries, hydrogen-powered forklifts can contribute to keeping the economies of African countries going for decades. And entrepreneurs can capitalize on opportunities to fuel, service, lease, and sell these forklifts.

Another promising opportunity for Africa is to utilize green and blue hydrogen to manufacture ammonia, which can then be used to make fertilizer. This is another pathway to economic diversification and industrialization. And, at the end of the day, we have to feed our people.

No Africans Left Behind

As we develop green and blue hydrogen projects in Africa, we must make sure Africans aren't left out in the cold. We should be taking steps now to ensure fair financing opportunities for African entrepreneurs interested in joining the hydrogen economy. And we must make sure African people get the training and opportunities they need to take hydrogen industry jobs at all levels.

Generally, Western companies are far more likely than their African counterparts to get financing. African funding sources exist, but many financial institutions are based in the West and feel more secure loaning money to western applicants. What's more, Africa is often perceived by financial analysts to be a risky environment—an unfair blanket assumption that robs Africans of opportunities in their own communities. Funding obstacles will be particularly problematic for entrepreneurs interested in hydrogen-related businesses because of the high costs of launching and operating them. It will be up to African leaders to develop fair financing policies and work with African entrepreneurs to compete for funding on a level playing field.

Our countries also need to look now at hydrogen industry local content policies. Like those for oil and gas, these policies should be fair to foreign companies but, at the same time, ensure that good-paying jobs, training opportunities, and contracts go to Africans and African businesses.

Morocco is a strong example of an African country working to build a hydrogen economy. Its all-hands-on-board approach, which includes cooperation among the government, corporations, non-profit organizations, and international organizations, is especially praiseworthy. In 2020, Morocco established a Hydrogen National Committee to oversee hydrogen feasibility studies and create a hydrogen roadmap. In addition, Morocco is working on creating infrastructure, looking into converting existing gas pipelines to hydrogen lines, and developing a logistics hub to ship hydrogen to Europe. The government is partnering with European countries, including Germany and Portugal, to develop hydrogen production research, while also exploring financing, research, and training programs.[20]

A Starring Role, Not a Supporting Part

In too many cases, throughout Africa's energy history, we've been our own worst enemies. Yes, other factors have worked against us when it comes to fully reaping the rewards of our natural resources; but ultimately, we're responsible for our own actions. We've dropped the ball too many times. We've mismanaged resources. We've focused on short-term gains as exporters instead of monetizing our natural gas value chains and building for the future. We've complained about foreign oil and gas majors but failed to negotiate good deals and develop policies that protect the interests of our people.

Now, we're moving toward a new era. Africa has an important role to play in meeting demand for no-carbon and low-carbon hydrogen. If we're not careful, we'll fall back into old patterns, behaviors, and choices that cost African people, businesses, and communities a chance at a brighter future.

We cannot let that happen.

Chapter 8

SOLAR POWER AND WIND: MORE WORK TO BE DONE

I MAKE NO SECRET of the fact that I'm a fan of oil and gas. I believe the industry has the potential to make poverty (and energy poverty) a thing of the past in Africa, not just by generating export revenues, but also by providing fuel to boost electrification. Electricity supports economic diversification and industrialization and improves standards of living.

I'm especially excited about the potential of gas. Africa's natural gas reserves are vast enough to meet the fuel and energy needs of hundreds of new thermal power plants (TPPs), industrial and manufacturing facilities, and residential distribution networks, while still serving export markets and existing supply systems. Even better, gas is a relatively low-carbon energy source that emits less in the way of greenhouse gas (GHG) than petroleum-derived fuels.

Gas remains an important part of Africa's future. At the same time, I'm also excited about renewable energy, including solar and wind power, and I believe Africa ought to develop its capabilities in these areas.

In the solar arena, we have an edge over other parts of the world. We get more sunlight than any other continent, and we ought to take advantage of that fact! In quantitative terms, Africa has 40% of the world's total theoretical solar energy potential.

This potential is enough to support the production of 60 million terawatt-hours (TWh) of electricity per year.[1] It's enough to keep 6,850 power plants with a generating capacity of 1,000 MW each operating at full nameplate capacity all day and night, 7 days a week, 365 days a year.

That's a staggering number. It's several orders of magnitude higher than current global production. According to the International Energy Agency (IEA), the world generated a total of 27.044 TWh of electricity in 2019.[2]

Although the numbers aren't quite as high as they are for solar, the potential for wind power in Africa is still substantial. According to a study published in 2020 by the International Finance Corporation (IFC), a division of the World Bank, Africa is technically capable of turning out 180,000 TWh per year of electricity using wind turbines.[3] This is 250 times the amount needed to meet the continent's current demand, and it's enough to keep at least 20 1,000-MW power plants operating at full capacity, day and night, 7 days a week, 365 days a year. (In practical terms, this is probably an overestimate, since the wind blows at varying speeds and sometimes doesn't blow at all. Nevertheless, the IFC study referenced here noted that two-thirds of Africa's wind potential is in places where local wind speeds average more than 7.5 meters per second, creating optimal conditions for wind turbines.)

But there's a catch.

Thus far, Africa's renewable energy potential has remained mostly in the theoretical realm. As of 2020, the continent had only 4,878.1 MW of installed solar generating capacity and about 6,000 MW of installed wind generating capacity.

So what's the source of this gap between theory and practice? And more importantly, *how* can we resolve it? African states can reap the benefits of solar and wind operations, but doing so will require a strategic regulatory framework, ongoing investments and cooperation from the international community, and ongoing efforts to strategically harness our natural gas resources.

Subsidies, Funding, and Other Considerations

There are several reasons why Africa is not yet in a position to put all of its renewable energy potential to work.

Solar and wind have an edge over fossil fuels in terms of operating expenses. That's because renewable facilities generate electricity using sunlight and wind (which cost nothing) instead of gas, petroleum products, or coal (all of which cost money) to generate electricity. These facilities are also less subject to wear and tear since they don't have to operate at high speed or high temperatures; therefore, they're less expensive to maintain. As the Union of Concerned Scientists has expressed: "Like most renewables, solar and wind are exceedingly cheap to operate—their 'fuel' is free, and maintenance is minimal—so the bulk of the expense comes from building the technology."[4]

Lazard, one of the world's top financial advisory and asset management firms, has developed an index known as the Levelized Cost of Electricity (LCOE) to measure just how much it costs to operate various types of power plants over their entire lifecycle.[5] This index—which takes geography, carbon emissions abatement, and other factors into account, as well as construction and operating costs—shows that solar and wind energy facilities are becoming increasingly competitive with gas-fired power plants.

It's worth noting, though, that the "other factors" used to calculate LCOE include various types of subsidies—such as preferential tariffs for renewable power plant operators, tax credits from local and national governments, public funding for research, and utility rebates and other incentives for consumers. These subsidies aren't the only reason why wind and solar power costs have gone down, and they are being phased out in some developed countries.[6]

However, subsidies aren't about to disappear. In fact, in its 2020 technical paper "Energy Subsidies: Evolution in the Global Energy Transformation to 2050," the International Renewable Energy Agency (IRENA) predicted that renewable energy projects were likely to continue receiving hundreds of millions of dollars each year in direct public-sector funding over the next three decades. It also indicated that it viewed these subsidies as necessary on some level to establish the resources and infrastructure to support the transition away from fossil fuels.[7]

The problem is that many African governments can't afford to subsidize renewables in this fashion. And they have reason to doubt that the rest of the world will fill the gap—either out of compassion, out of concern for the climate, or on principle. After all, members of the United Nations Framework Convention on Climate Change (UNFCCC) have been saying since 2009 that the world's most economically developed countries were committed to making USD100 billion per year available to developing countries by 2020 for the purpose of mitigating climate change, and they have not kept that promise.[8]

Skill and Technology Deficits Could Leave Africans in the Cold

Let's be honest: Africa needs more than the ability to provide renewable subsidies. We need the kinds of investments—and cooperative relationships—that empower the average African to fully capitalize on renewable energy-related opportunities.

Yes, we want to use wind and solar solutions to deliver electricity to more Africans. Yes, the presence of reliable power in and of itself will help Africans build a pathway to a more stable, prosperous future. But that doesn't mean we should be content to let the lion's share of wind and solar energy jobs—particularly high-paying technical, professional, and leadership positions—go to non-Africans. And that doesn't mean that we're OK with most of the contracts and business opportunities going to non-African companies and entrepreneurs.

IRENA estimated in 2020 that renewables will create 45 million jobs around the world by 2050.[9] My question is: How many of those jobs will go to Africans? As things stand, likely only a fraction of them will. When we talk about growing Africa's renewables industry, we cannot ignore the significant capacity deficits that exist throughout the continent. In 2018, the World Bank reported that while solar power was a promising means of addressing energy poverty in Africa and Asia, both continents lacked "job-ready talent to finance, develop, install, operate and manage solar power systems and other-off-grid energy solutions."[10]

Even one of the continent's most industrialized countries, South Africa, has considerable room for improvement in this area. African human resources firm AltGen in 2021 noted that recruiting South Africans for technical solar and wind positions is a struggle, and filling some positions takes months.

"There is a general scarcity of experienced technical support skills in South Africa's PV space [PV or photovoltaic cells absorb sunlight and use the light energy to create an electrical current] such as electrical engineers, operations and maintenance managers, and mechanical technicians," AltGen wrote. "While most entry-level engineers graduate from universities are equipped with a bachelor of engineering or a bachelor of science in either electrical or mechanical engineering, most of them would have little to no exposure to renewables and therefore cannot add any form of substantive value from day one. It then falls on the employer to get them into some type of learnerships or mentoring programmes to facilitate skills transfer needed to equip them with relevant industry skills like solar PV, wind, energy storage, etc. depending on the industry, and technology."[11]

AltGen, at least, has responded to this skill gap by developing a Renewable Energy Specific Skills Programme for independent power producers (IPPs) and operation and maintenance (O&M) providers to help residents in South Africa's Northern Cape find long-term employment. I hope others follow their example. But developing local capacity must be a full-steam-ahead kind of effort throughout the continent.

One of the benefits of our oil and gas operations is that, as a result of local content policies, they support capacity building: Africans acquire entry-level positions but they also receive training for skilled positions. Africans fill professional and leadership roles in oil and gas, and they run their own companies in the industry (or in roles that support it). African countries have spent decades developing and fine-tuning local content policies specifically designed to make sure that foreign operations don't reap most of the economic opportunities their petroleum resources represent and leave Africans with breadcrumbs.

As we begin embracing solar and wind, we must act to make sure workers from North America, Europe, and Asia don't sweep in to "help" poor Africans develop the facilities they need—while "helping" themselves to good-paying jobs and business opportunities that should go to Africans. African governments should be working now to develop and fine-tune renewable energy policies with a strong local content component. We must insist on technology and skills-sharing agreements among Western investors and Africans. We must work to develop financing initiatives so that local entrepreneurs can compete effectively with their Western counterparts.

Unless we proactively protect Africa's interests, Africans will pay the price in lost jobs and missed economic opportunities. Even in areas where African independent power producers (IPPs) have entered the renewable energy sector, these companies will likely struggle to compete with their foreign counterparts. In fact, it's already happening. In 2017, for example, about 10 Tanzanian-owned solar home system

(SHS) providers were put out of business by foreign SHS companies that took advantage of grant money from Western institutions to push down their prices. I've seen this infuriating pattern repeat itself throughout my travels around Africa.

Africans deserve a level playing field so they aren't left out of opportunities in their own countries. We need more local content regulations geared for renewable energy sectors, capacity-building programs, and financing initiatives for local entrepreneurs.

The Challenges of Intermittency and Storage

Africa has huge potential for solar and wind power generation. Nevertheless, there are still times when the sun doesn't shine because the sky is covered with clouds—or simply because night has fallen. Likewise, sometimes the wind doesn't blow (and it certainly doesn't blow at the same speed everywhere). At such times, renewable energy facilities cannot sustain electricity production at their nameplate capacity.

This is a problem since electricity can't be stored easily for use later, as is the case with tank farms for hydrocarbon-based fuels or silos for coal. Of course, it can charge high-capacity batteries capable of covering some of the gaps that emerge when solar panels and wind turbines can't operate. However, these batteries have a number of disadvantages, including but not limited to the following:

- **High cost:** The cost of a battery capable of providing residential or business users with a back-up capacity of just 8 kW (equivalent to 0.0008% of the capacity of a 1,000-MW TPP) comes to around USD5,000-10,000.[12] There are larger batteries available, but they're no less costly. For example, the 0.8-MW Megapack unit sold by Tesla carries a price tag of more than USD1.2 million.[13] These rates are more than many African organizations can afford to pay.

- **Limited capacity:** Battery storage can make homes and other structures that depend on solar panels for electricity more resilient, but it can't keep them operating at full capacity for more than a few hours. This is especially true for facilities that depend on power-hungry HVAC (heating, ventilation, and air conditioning) systems. What might happen, for example, if data centers or refrigerated warehouses can't be sure if their cooling units will stay on all night? The consequences of unexpected temperature fluctuations might be devastating.
- **Sourcing issues:** So far, there are no facilities in Africa where market-ready lithium-ion batteries (the most common type of batteries used to store electric power) are manufactured.[14] There are plans to establish such facilities in South Africa, but as of this writing, they have not yet come to fruition.[15] As a result, most African countries aren't yet part of the value chain for the solar power sector—and this means that investments in African solar power projects may not create many jobs for Africans.
- **Concerns about cobalt:** One African state that has come to play a key role in the value chain for renewable energy—the Democratic Republic of Congo (DRC). The DRC is home to more than half the world's cobalt reserves, which is important because this metal is critical to the production of lithium-ion batteries; however, it's not always mined responsibly or sustainably. According to information gathered by CBS News and Amnesty International, there are tens of thousands of child laborers involved in cobalt mining in the DRC. Some of these "miners" are as young as four years old, and most of them have never been to school. What's more, these children—and the adults who work alongside them—have little protection from the toxic fumes they encounter while handling cobalt ores.[16][17] This is simply unacceptable. These laborers deserve better!

Mini-Grids

As I hope I've made clear thus far, I believe Africa should take a gas-first approach to alleviating energy poverty. But I want to acknowledge the arguments in favor of mini-grids: small, localized power generation and distribution systems that use renewable energy sources to produce electricity for rural communities that can't easily be connected to national or regional transmission grids.

Mini-grids have obvious appeal. They meet isolated communities' need for energy without contributing to carbon emissions, and they can be built without any help from state-run power companies, which may be unreliable, corrupt, or simply lacking in resources. They set a precedent for establishing energy systems that never depend on fossil fuels at all.

For these reasons, mini-grids ought to be built wherever they're feasible. But the challenges that mini-grids face are real, and they shouldn't be overlooked. I'll mention a few here:

Policy: Most African countries haven't established a policy environment that can support mini-grids. They haven't streamlined the permitting process,[18] and they haven't established tariff policies that are transparent enough to reassure commercial investors.[19] Nor have they adopted legislation or regulations that prevent government-owned entities from expropriating or stranding mini-grids in the event that they expand national grids into the same areas that were previously served by small, independent operators. They haven't met the goals outlined by Sustainable Energy for All (SEforALL), which said in 2020: "In an ideal policy and regulatory scenario, mini-grid owners would have reassurances that, should grid arrival occur, a range of options would be available to them, such as receiving compensation and operating alongside the main grid."[20]

Money: Mini-grids are not exempt from the reality that renewable energy projects face high start-up capital costs and high storage costs. (Solar panels, wind turbines, and batteries are expensive, even if sunlight and wind are free.) They're also unlikely to turn a steady profit, SEforALL noted, since many of the customers that they aim to serve will have difficulty paying for electricity supplies due to poverty, as well as seasonal and weather-related fluctuations in farmers' incomes.[21] As a result, mini-grids will have a hard time securing commercial financing without some outside financial support, whether in the form of a government subsidy or foreign contributions. As Jack Holmes, the Vice President of UK-based Winch Energy, told the World Bank's International Finance Corporation (IFC) in mid-2020, "Few utilities in Africa are profitable. When it comes to off-grid [electricity], you're going to need the same amount of state support or donor subsidies that the on-grid has."[22]

Local capacity: Renewable energy infrastructure is not always easy to maintain or repair, especially in isolated communities that are difficult for technicians to access. Additionally, the regions of Africa with the highest renewable energy capacity may also experience climate conditions that shorten the service life of renewable generating facilities.

"It is one thing to install a renewable system in a remote community and a whole other to ensure its long-term, sustainable operation," Feyza G. Sahinyazan, Beedie School of Business, Department of Technology & Operations Management, Simon Fraser University, and Serasu Duran,

Operations and Supply Chain Management at Haskayne School of Business, University of Calgary, wrote for *The Conversation.*[23] "A lack of community involvement and governance can mean efforts go to waste, as our research shows."

In 2021, Sahinyazan and Duran completed an investigation of renewable mini-grid projects in rural areas during a 10-year period. They found that a significant number of the projects failed to live up to expectations.

"In one particular context, 60 percent of renewable mini-grid projects were abandoned within just six months of installation. The reasons cited for failure always point to the same challenges: an absence of local maintenance expertise and a lack of acceptance" Sahinyazan and Duran wrote.

The above points aren't a comprehensive list of the hurdles that investors in mini-grids will have to surmount. But it should serve to illustrate that these projects aren't a panacea for energy poverty in Africa's rural communities.

Certainly, the idea of mini-grids shouldn't be abandoned. If they can be built feasibly, they should be built. But if they can't, there's an alternative: gas. I don't mean natural gas, but rather liquified petroleum gas (LPG), which can be produced from natural gas. I've already mentioned LPG as a fuel for cooking, but I'd like to point out that it can also power generators—and it can do so more cleanly than gasoline or diesel.

Gas First. Solar and Wind Later.

So, what's the best way to address the challenges of implementing solar and wind power in Africa? I believe Africa should start with gas.

More specifically, African countries that possess natural gas should develop their reserves and do everything possible to create domestic gas markets. They should use their gas to meet domestic energy demand in full—by generating electricity, by supplying fuel to large industrial facilities, and by making LPG available to household consumers.

Then they should move on to sharing with their neighbors, if they can. If they have more than they need, they should look into exporting surplus gas (or surplus production from gas-fired TPPs) to nearby countries. Doing so will help ensure that gas-producing states aren't the only ones who can count on having enough fuel and power. It will create incentives for cross-border cooperation and reduce dependence on dirtier fuels such as coal and petroleum products. It may also lay the foundation for regional energy markets capable of serving many African countries at once.

As this process moves forward, African countries will be in a position to install more and more solar panels and wind turbines—not as primary sources of power or as alternatives to fossil fuels, but as complements to the gas-burning TPPs that make sure the lights can always stay on.

With enough time (and enough gas), the hurdles Africa faces with respect to renewable energy will shrink. Eventually, African states will be able to produce the electricity they need.

With cheap and abundant energy, these countries will be in a better position to create new jobs and sustain economic growth. They will become wealthier and more developed. Their citizens will be able to expand their focus beyond short-term survival and exercise more freedom to contemplate long-term goals such as reducing carbon emissions and demanding better treatment for mine workers. At the same time, they'll be able to buy more of the expensive goods and equipment they need to establish solar farms with back-up battery storage.

And along the way, African countries should also find ways to stake out their positions along the renewable energy value chain. They should, for example, focus on extracting natural resources such as cobalt while also developing their reserves of rare earths, which can be found in multiple African countries.[24] (Rare earths are used in solar panels and wind turbine magnets.) They should try to boost the production of copper, which is suitable for use in electrical wiring, and is abundant in the DRC, the world's fourth-largest copper producer as of 2022.[25] Additionally, they should try to attract investors to their deposits of iron ores, along with minerals such as chromite, manganese, nickel, tantalite, and tungsten, which can be alloyed with iron to produce specialty steels.[26]

However, African countries should also look for opportunities to establish processing and fabrication facilities capable of putting these resources to use. Whenever and wherever possible, elements mined in Africa should be used *in Africa* to make solar panels, wind turbines, batteries, and other components instead of being transported halfway around the world and back.

Gas can even play a role in helping Africa secure other links in the renewable energy value chain. That is, gas-producing countries can choose to direct some of their output to petrochemical plants that specialize in the production of the plastics and resins utilized in turbine blades.

In short, Africa needs more than access to solar and wind power. We need solar farms and wind parks to manufacture some of the components used in such facilities, and to source some of the materials that form part of these components. So build solar farms. Build wind parks, too. But build them in Africa! Let them create jobs for Africans—both in the mines and at the manufacturing and generating sites.

But above all, Africa needs to build up the gas sector first. We should recognize the potential of gas to give Africa what it needs to evolve and support healthy renewables industries—and bring that potential to life.

Chapter 9

FINANCING A JUST ENERGY TRANSITION: OUR PATH TO OVERCOMING OBSTACLES, INTERFERENCE, AND UNKEPT PROMISES

At the Paris Climate Conference in 2015, industrialized nations, eager to entice developing countries to join them on the march toward net-zero, offered the usual inducement of the rich and powerful to those with less: USD100 billion per year until 2025 to facilitate a smooth transition.[1] The money could be spent in any way the recipient countries saw fit: They could invest in adopting new, green technology or mitigate some of the harm already caused by global warming.

This offer sounds like a good incentive on paper, unless you consider that those wealthy countries had fueled their progress with oil and gas, and then they were jockeying to deny less-developed countries the same privilege—and attempting to force those countries to make the transition on a compressed timeline.

What's more, this offer barely scraped the surface of how much the world will have to spend to keep global warming below 2° C: It has been estimated that achieving global net-zero by 2050 will cost as much as USD130 trillion.[2] For developing countries, the cost simply to adapt to the current effects of climate change will be USD70 billion per year, according to the UN.[3] In 2020 alone, global spending on the energy transition totaled more than USD500 billion, according to BloombergNEF's (BNEF's) *Energy Transition Investment Trends* report.[4]

Perhaps the most problematic part of this incentive program, however, was in its execution. A half-dozen years later, just ahead of the 2021 Glasgow Climate Summit, people began asking: How well had those developed countries delivered on their pledge?

The answer: not very well at all.

In 2018, the payout—some of it grants and gifts, much of it loans—was about USD78 billion, a shortfall of nearly 25%. A year later, the sum had risen slightly to USD80 billion, but it retreated in 2020.[5] By the way, those figures are based on self-reporting to the Organisation of Economic Co-operation and Development (OECD), which comprises mostly wealthy nations motivated to look good to the international community.

Granted, USD80 billion isn't exactly pocket change. Still, it's well below the promised total, which was insufficient to begin with. And though the wealthy countries assembling in Glasgow vowed to do better, how they would make up for their deficits was left unsaid. No wonder skeptics scoffed.

"Making good on a promise made (years ago) is setting a pretty low bar for a successful COP26," Teresa Anderson, climate policy coordinator at non-governmental group ActionAid International, said in a statement reported by *The New York Times*.[6]

So how did this promise of support become little more than a goodwill gesture?

The problem, it seems, stems from bad planning and inadequate fundraising.

Funding Roadblocks and Misleading Signposts

Getting the money together was the first hurdle. None of the participating countries involved was ready to simply hand out cash in million-dollar increments. No country was willing to foot more of the bill than it thought was fair. For that matter, no one had actually committed to paying a specific amount. Each country simply agreed to pay something, in the hopes of getting the next country on board.

So how did they actually intend to pay for their promises? In general, each nation would have to raise the money on the open market, something they were optimistic they could do. They'd look to institutional investors, pension funds, and the like. With climate activists influencing investment decisions, this seemed like a safe bet. The only problem was that the same funds that would gladly back wind turbines and solar panels in the U.S. and Europe were less enthusiastic about making similar investments in other parts of the world. Even when the rich nations extended unique incentives, including a willingness to swallow any losses, they didn't get much traction.

And then, of course, came the question of the amount of money in question: As it turns out, nobody seemed to know exactly how much money was raised in the first place.

Critics say the numbers the OECD has disclosed were the result of misleading or downright false reporting. For example, Oxfam International, which fights global poverty and injustice from its headquarters in Nairobi, Kenya, calls the OECD's figures inflated, and by much more than a penny here or a euro there. By Oxfam's estimation, the climate financing total for 2017-2018 was closer to USD22.5 billion,[7] a fraction of the "official" numbers. (Of course, some might say Oxfam is engaging in their own form of obfuscation: Although their figure includes grants, it only considers the amount of loans made at below-market rates.)

As significant as those charges are, perhaps equally damning is the charity's claim that some countries are describing aid as "climate relevant" when it is nothing of the sort. When funds used to build roads are reported in the climate mitigation total, that's some suspect accounting, indeed.

Missing the Mark

Obviously, some money is flowing to its intended purpose. Most of it is being transferred directly from one country to another in the form of public grants or loans, with multilateral development banks (MDBs) and private finance making up the difference. This, too, warrants a caveat: Most of the money that has changed hands has gone to reduce greenhouse gas (GHG) emissions, which are easier to measure than the results of more complex programs aimed at supporting adaptation to climate events like increased flooding, extreme heat, and more powerful storms.

But there's an even bigger issue. Not enough dollars are going where they're needed the most. Global consulting giant PricewaterhouseCoopers (PwC) noted the allocation to Africa falls significantly short of what the continent requires to meet global emissions targets.[8]

The continent's vulnerability to the effects of climate change is well recognized. The shortfall reflects an issue of access, or rather, the lack of it. The fact is, middle-income countries that are better equipped to apply for funding are receiving the lion's share of distributions while African nations that lack the know-how are often overlooked.

"Many, many African countries are lamenting that they are not able to jump through the hoops (to access climate finance) because of the complexity and the technicality," Chukwumerije Okereke, an economist at Alex-Ekwueme Federal University Ndufu-Alike in Ikwo, Nigeria, told nature.com. "And they're not receiving sufficient capacity-building exercises and training in this."

As nature.com noted, sometimes money headed for the poorest countries never makes it there. The website said that when the International Institute for Environment and Development tried to track funding for adaptation projects in the UN's least developed countries—33 of them in Africa, from Angola to Zambia[9]—they found they could account for just 20% of the USD30 billion they received.[10] No one could say for certain whether it was diverted, lost, or stolen, but it isn't going through the proper channels or getting into the appropriate hands.

Worse yet, the West is clamping down on investment in African fossil fuels, cutting off the continent's capacity to pay for at least part of its own energy transition.

These are a few more reasons why Africa needs to take charge of its own energy. When Western countries get involved, the result is often a well-intentioned path, paved with potholes!

Critical Questions

With developed nations unable to honor their USD100 billion promise, I must ask: Is it fair to expect poorer countries in Africa and elsewhere to keep their end of the bargain, bowing to unrelenting pressure to reduce their GHG emissions—especially when the amount they contribute is so small and getting to net-zero is so expensive?

Here is something to consider: PwC has estimated it will likely take a USD2.8 trillion investment for Africa to transition from its current energy base by 2050.[11]

According to PwC, "27.32 exajoules of additional renewable generation would be needed in Africa's energy mix to compensate for the declining use of fossil fuel." Given that the current renewable generation is just 1.79 exajoules, that's an unimaginably large increase.

Another thing to consider: the International Renewable Energy Agency's (IRENA) *Scaling Up Renewable Energy Deployment in Africa* report shows that Africa has the potential to install 310 gigawatts of clean renewable power—or half the continent's total electricity generation capacity—to meet nearly a quarter of its energy needs by 2030. Getting there will require an annual investment of USD70 billion, however. (Adding insult to injury, IRENA says that while renewables will create 45 million jobs around the world by 2050 and lead to a 2.4% rise in the global GDP, Africa and Asia lack "job-ready talent to finance, develop, install, operate and manage solar power systems and other off-grid energy solutions."[12])

The multi-trillion-dollar question is this: how can Africa pay for the transition and meet a sustainable level of renewable capacity if a financing source such as oil and gas proceeds disappears and no one else is pitching in? I believe there are reasonable and intelligent ways forward. Like most things that are worth doing, however, it will require some heavy lifting on Africa's part.

Successes We Can Build Upon

First, if we are going to attract backing for renewables projects, we must continue to create an enabling environment.

Thankfully, we've come a long way from the corrupt business practices that often accompanied deal-making a generation ago (if not more recently). Many countries have made considerable strides in creating an honest and investor-friendly climate.

Take South Africa, for example. It's the world's 12th largest emitter of greenhouse gases,[13] meaning there's lots of room for improvement. But it's also 12th on the list of the most attractive countries for renewable investment,[14] largely due to its Renewable Energy Independent Power Producer Procurement Programme (REIPPPP).

For more than a decade, REIPPPP has successfully encouraged direct private investment in solar, wind, biomass, and small hydro projects, attracting more than R120 billion in 2014.[15]

The program allows independent power producers to design, develop, and operate renewable energy power plants across the country, and it has succeeded in no small part because of its transparent procurement and implementation framework.

By 2018, REIPPPP had procured 102 independent power projects from four bidding rounds.[16] Among them were the De Aar Solar Power and Droogfontein Solar Power plants, which were built in 2014 at a combined cost of R3 billion. Since then, Globeleq South Africa, the lead investor in the two plants, has added seven more majority-owned and -operated wind and solar assets for a total capacity of 450MW; it will build six wind and solar PV projects as the result of the fifth bidding round. That includes the first renewable projects of any kind in the Free State province as part of the REIPPP Programme. Once these new projects are commissioned, Globeleq South Africa will have added a total generation capacity of 1,274 MW to the grid.

In 2021, South Africa committed to procuring up to R208.5billion in renewable energy under the REIPPPP.[17]

In a world where success breeds success, it's no surprise that South Africa has caught the eye of investors. In an interview with African Business, Bhavtik Vallabhjee, head of Power, Utilities, and Infrastructure at South Africa's ABSA Group, said the country had been the "low hanging fruit for financial institutions over the last few years due to the large volume of transactions over a short period of time."[18] In other words, one good thing has led to many more.

Of course, it also helps that there are a handful of organizations ensuring regulatory certainty, including the South African Power Pool (SAPP) the Regional Energy Regulators' Association of Southern Africa (RERA), and the Southern Africa Development Community Centre for Renewable Energy and Energy Efficiency (SACREEE). As Obbie Banda, Underwriter at the African Trade Insurance Agency (ATI), told *African Business*, "One of the biggest challenges for project developers focusing on renewables in Sub-Saharan Africa is that the regulations in some countries have never been clear."[19]

It's also unsurprising that the South African model has inspired imitators.

In Uganda, for example, the Global Energy Transfer Feed-in Tariff (GET FiT) is a standardized, competitive program that enables private developers to fund small-scale renewable energy generation projects—and ensures they will be able to sell the energy they produce to the grid through offtake agreements, overcoming a concern that has often hindered investment. The program began with a small portfolio—17 facilities with a total installed capacity of about 160MW. According to *African Business*, GET FiT has now attracted USD450 million in investments that should supply enough energy for about 200,000 households. Similarly, Zambia has adopted the GET FiT strategy, and Ethiopia isn't far behind.

We Can, and Must, Be Part of the Solution

Bringing Western investment to our energy projects is meaningful, and capital from foreign investors is always welcome. But it could also inevitably place us in the same scenario Africa has been chafing against for generations: outsiders telling us what's best for us when really, it's more about what's best for them.

I've made my position on this very clear: It's time we wrest ourselves from our overreliance on foreign involvement. Yes, we must continue to make it easy to do business with us. We need to be more transparent, to fast-track projects, to cut our bureaucracy—but we should no longer beg for aid that will be awarded on the condition that we abandon our homegrown oil and gas assets and aspirations. Instead, we must depend more on ourselves, not others, to ensure a just transition. We must find ways to bring energy to all of our people, whether from natural gas or hydrogen or solar. We must create a transition that considers the unique needs and priorities of African countries.

And we must figure out how to pay for it ourselves.

As Senegal's President Macky Sall, chair of the African Union, said, African states are open to embracing renewable energy sources: The problem comes when we are bullied into giving up our fossil fuels, and the opportunities they represent, on others' timetables. "Our countries cannot achieve an energy transition and abandon the polluting patterns of the industrialized countries without a viable, fair and equitable alternative," Sall said. "Our countries, which are already shouldering the crushing weight of unequal trade, cannot bear the burden of an unfair energy transition."

South African President Cyril Ramaphosa echoed those thoughts. "We must contend not only with these primary dangers (of climate change), but also with potential economic and social damage should the global community fail to deal with the crisis in a way that works for developing as well as developed markets," he said.

So how do we achieve a fair and just solution? I say, it's time to develop African energy banks that will advance fossil fuel development and infrastructure while also supporting low-carbon initiatives. In addition to achieving these critical objectives, we also will send a clear signal to the marketplace that Africans can become leaders in scaling up private capital.

We can find Inspiration in the African Export Import Bank (Afreximbank). Founded in 1993 through a consortium of African governments and public and private investors, Afreximbank finances and promotes intra-African and extra-African trade. Year after year, the institution fulfills its mission and meets its goals. Between 2015 and 2019 alone, Afreximbank disbursed more than USD30 billion in support of African trade, including more than USD15 billion for the financing and promotion of intra-Africa trade.[20] In 2020, Afreximbank received the Africa-America Institute's (AAI's)

Institutional Institution of Excellence Award for its commitment to the creation and implementation of the African Continental Free Trade Agreement (AfCFTA) and its ongoing dedication to investing in education.

Afreximbank, by the way, recognizes the importance of protecting Africa's oil and gas industry. "The way we see it at the bank, Africa produces less than 4% of greenhouse gas. We are not the problem of greenhouse gases. We are the victims. We are asking for balance," said Afreximbank President and Chairman of the Board Benedict Oramah.[21]

With the support of one or more African energy banks, local oil and gas companies will be able to acquire assets and expand operations. These funds can finance building pipelines and upgrading refineries to produce cleaner fuels and drive industrialization. African banks can expand the work of Mali, Namibia, Gambia, Senegal, Mauritania, Niger, and South Africa, which are already on the leading edge of the hydrogen economy through their emerging green hydrogen operations.

Africa's Way Forward

An African energy bank is more than an idea or long-term goal; it's already in the early stages of development. In May 2022, Afreximbank signed a Memorandum of Understanding (MoU) with the African Petroleum Producers Organization (APPO) for the creation of a multi-billion-dollar energy bank.[22]

I can almost hear you wondering how this can work. How can a continent that is struggling to bring many of its people out of poverty raise the capital required to establish a bank and begin loaning money?

We can start by partnering with investors here and in the diaspora. As of December 2020, total private wealth in Africa alone totaled approximately USD2 trillion. When the World Economic

Forum said it was time for wealth—instead of poverty—to dictate Africa's development,[23] it meant this: It's time for Africans to invest in the solutions so Africa can progress and loosen the ties of foreign intervention.

We can also leverage the royalties that oil and gas companies pay to African governments. The African Energy Chamber projected that governments would earn a tidy USD100 billion in 2021 from royalties, profit oil—that is, the amount of production, after costs, divided between partners in a production-sharing contract—and other taxes. Granted, governments are not going to turn over all of their oil and gas income; they depend on it too much for the programs that keep their countries going. Yet even 5% of that amount would produce USD50 million in seed money for a bank dedicated to financing energy exploration, development, and infrastructure.[24]

Most of all, those funds would be a symbol that Africa can do things for itself—and do just fine without broken billion-dollar promises.

The Potential of Pension Funds

Direct investment in oil and gas projects through pension funds is another option for raising capital to finance energy projects. According to Capetown-based investment firm RisCura, local pension funds collectively manage around USD350 billion of assets in Sub-Saharan Africa, and they are actively looking for new places to invest.[25] Why not encourage them to add oil, gas, and renewables projects to their list?

Investing pensions in the energy sector is hardly a new practice. Pension funds around the globe are investing in green energy projects, and some of America's largest pension funds are invested in fossil fuel producers. Natural gas investments, in particular, offer significant opportunities for return. This is addressed in detail in the Africa Energy Chamber's Q1 2022 Outlook, *The State of African Energy*, which predicts strong long-term global demand for natural gas and LNG.[26]

Chapter 10

AFRICA'S OIL AND GAS INDUSTRY NEEDS GOOD GOVERNANCE MORE THAN EVER

2021 WAS A historic year for the African oil and gas industry. On August 16, 2021, Nigerian President Muhammadu Buhari signed the long-awaited Nigerian Petroleum Industry Bill (PIB). In the making for some 20 years,[12] the PIB will bring urgently needed reform to Nigeria's petroleum industry and ensure fairer, more streamlined, and more profitable use of its oil resources.[3]

The journey to this milestone saw numerous stops, starts, and detours before Nigeria's House of Representatives and Senate passed the PIB in July 2021, and the President signed it. During the last two decades, multiple governments attempted without success to pass the bill, and conventional wisdom claimed it was destined to fail because of its scope and complexity.

I believe the PIB will transform Nigeria's oil and gas industry and, ultimately, the country as a whole. It promises to address the inefficiencies plaguing the Nigerian National Petroleum Corporation

(NNPC), from slow approval for oil projects to budget shortfalls that hinder its ability to pursue public-private partnerships. The bill is intended to create a supportive environment for both international oil companies (IOCs) and indigenous firms, protect the environment and the interests of host communities, support economic diversification in Nigeria, and—crucially—promote more transparency in the administration of petroleum resources.

Promising components of the PIB include its emphasis on giving back to African communities. Oil and gas exploration can have lasting effects on local communities, from environmental damage to a negative impact on fishing, agriculture, and other sectors people rely on for their livelihoods. This has been particularly true in the Niger Delta. The PIB takes a step towards addressing those concerns by requiring upstream companies to contribute 3% and other companies to contribute 2% of their operating expenditures from the preceding calendar year to the host communities' development trust fund.[4]

Also exciting, the bill requires gas-producing licensees or lessees to submit an elimination and monetization plan for the gas generated during oil exploration and production. In other words, Nigeria finally will capitalize upon its gas to promote economic growth and business opportunities, instead of wasting this valuable resource and releasing greenhouse gases during flaring. These articles of the bill complement Nigeria's Decade of Gas Initiative, which calls for enhancing gas monetization through large-scale projects, including gas-to-power developments.

Additional provisions include:

- Incorporating NNPC as a limited liability company with the aim of strengthening accountability and transparency.
- Establishing the Nigerian Upstream Regulatory Commission as a technical regulator.

- Establishing the Nigerian Midstream and Downstream Petroleum Regulatory Authority, replacing the Petroleum Products Pricing Regulatory Agency (PPPRA) and other downstream and midstream regulators.
- Lowering royalties and tax rates and improving terms for marginal fields and local companies.
- Establishing the Frontier Exploration Fund to direct 30% of NNPC limited profit to the exploration of unassigned frontier basins.
- Introducing dispute resolution mechanisms between governments and oil companies.

The passage of the PIB is proof that change is possible. It represents a significant political victory for President Buhari and sends a strong message to international investors. As Africa and the world adapt to the post-COVID economy, the bill is a critical and overdue step toward capturing more of Africa's oil revenue and boosting federal budgets.

The PIB also serves as a reminder of both the best and worst aspects of our industry: It proves we have the power to harness our resources effectively and shape our future. But it should never have taken 20 years to see the bill signed. In that time, we've lost a lot of ground. We're not keeping up with the pace of change globally.

Legislation like this is more important than ever. Our industry is at a crossroads: The challenge of staying competitive is increasing, as is the pressure to move away from fossil fuel use and production and toward renewables. Funding for the exploration of fossil fuels is being cut. Money is drying up.

What's more, many large IOCs are beginning to divest in Africa because of the issues emerging around the energy transition. Royal Dutch Shell and ENI, for example, are seeking to sell their oil and gas operations in Tunisia. Some are re-focusing on North American shale exploration. Some are re-focusing on green energy.

Just as the continent's governments must take steps to ensure an "Africa First" energy transition, one with Africans' needs in mind, they must also continue to ensure a thriving, strategically managed, "Africa First" oil and gas industry. If Africans are going to realize the benefits of our natural resources, our governments can't drop the ball. That means government reform is critical, from retooling ineffective oil and gas regulations to getting serious about eliminating the scourge of corruption.

But the work doesn't end there: The post-COVID environment in Africa—with the accompanying challenges related to the decline in oil prices and a slowdown in operations—has created a perfect storm of opportunity. We can step into the void and meet the needs of our industry and our partners by actively developing fiscal policies that make oil and gas operations in Africa more profitable and sustainable. We need to provide compelling incentives during the current downturn. We have to create fiscal terms that reflect the current market reality, especially in the wake of the pandemic. And we need to incentivize exploration, institute fairer tax policies, and cut the red tape.

It's our responsibility to build, educate, and mobilize the energy sector for our own growth and prosperity. We do that by achieving a stable, transparent, business-friendly environment. I've been vocal about the need to adopt free market policies to attract capital. To the extent we can offer regulatory and tax relief, pro-business policies, and more streamlined processes, we can create the enabling investment environment that will drive our growth. That is the ticket to a successful energy transition and a sustainable future.

But we must move faster. Our failure to move more quickly and decisively on legislation like the PIB is one of the biggest challenges we face. The slow pace of change has plagued our industry for decades, and it's one of the biggest obstacles to our progress.

Why Are We Playing the Waiting Game?

Africa's approaching energy transition is a reality. That puts us on a tight, unforgiving deadline for fully harnessing our oil and gas industry. We have to ask ourselves: Why, when the wheels of the energy transition are spinning faster, are we still moving so slowly? We need to accelerate the pace of negotiations, and governments need to make prompt licensing round awards. It's not surprising that investors are reluctant to invest in new projects when our approval process takes so long.

Ghana and Uganda are good illustrations of the delays in implementing energy projects. Both countries discovered oil in 2006, and Ghana started producing oil from its Jubilee Field discovery just three years later. Uganda, by contrast, is still waiting. Tullow Oil, the company that made the initial finds in the Lake Albert region, experienced multiple setbacks and decided to sell to Total (now TotalEnergies) for USD575 million in early 2020; however, the French major didn't make a final investment decision (FID) on the Lake Albert fields until April 2021, after finally resolving Tullow's long-standing tax disputes with the Ugandan government.

It's great news that this project is finally moving forward, putting Uganda on the path to first oil by 2025. Remember, though, that the fields in question were discovered in 2006. That's a wait of nearly 15 years, and that's simply not acceptable. African projects are not attractive investment targets if it takes that long to secure the necessary approvals. The industry is moving and changing too fast globally to accommodate foot-dragging by African countries, and we have a host of new realities to contend with.

Growing Competition from Outside Africa

One of those new realities—and another pressing reason we need to move faster—is that our competition is changing. The biggest competition in Africa is no longer between Uganda and Ghana, or Egypt and South Africa. We're also competing with two small countries in the Caribbean: Guyana and Suriname.

The Guyana-Suriname basin is now attracting more new investment than many other established oil provinces, including the Gulf of Mexico. Why? It's partly because of the amount of oil and gas involved. The U.S. Geological Survey has estimated that the basin holds more than 13 billion barrels of oil equivalent (boe). But it's also because these countries are doing a better job of creating a favorable business environment for oil and gas.

In Guyana, ExxonMobil has already started production at a section of the Stabroek block, Liza-1, where it hopes to extract up to 120,000 barrels per day (bpd). The U.S.-based giant is due to bring the second section of the block—Liza-2, which will yield 220,000 bpd—in 2022 and will follow with the 220,000 bpd Payara section in 2024 and the 250,000 bpd Yellowtail section in 2025. ExxonMobil is also looking at other parts of the block with the intent of launching at least two more development projects before the end of the decade. Altogether, Stabroek may see output levels reach one million bpd by 2027.

ExxonMobil isn't able to do this just because Stabroek contains huge reserves—more than nine billion barrels, according to the most recent estimate. It also benefits from the production-sharing agreement (PSA) it signed with Guyana, which included favorable royalty and profit-sharing terms that will make the project profitable even if crude prices remain weak. The contract has drawn no small amount of criticism within Guyana for being too generous to ExxonMobil. Even so, the government wants work to proceed on this project, which is expected to push Guyana's real GDP to expand by 47.95% in 2022.[5]

Meanwhile, in Suriname, ExxonMobil, Royal Dutch Shell, TotalEnergies, Apache, and several other companies are all preparing for major operations off Suriname's coast. Apache and TotalEnergies made 3 huge discoveries at the 566,559-hectare (1.4-million-acre)

Block 58 in 2020 (Maka-Central-1, Sapakara West-1 and Kwaskwasi-1) and then found another large field (Keskesi-1) in 2021. Apache is so convinced of this region's potential that it has dropped its Permian Basin rigs in Texas to concentrate on Suriname.

What's making Suriname's offshore fields even more attractive is the producer-friendly business environment that President Chan Santokhi has been fostering since he took office last year. Under his leadership, Suriname has made a practice of signing 30-year PSAs, offering terms that are considerably longer than those of neighboring countries. This gives oil companies more time to ramp up their operations and also to continue drilling. And since overall production costs are relatively low, these favorable terms make it even easier for upstream operators to turn a profit, even if oil trades in the USD30-40 per barrel range.

This is the kind of strategy we ought to follow to attract IOCs. We can't afford to lose revenue to countries that are doing a better job of creating an enabling environment for oil and gas investment.

What might that look like in African countries? Consider Gabon.

Reforms: A Step in the Right Direction for Gabon

Gabon's adoption of a new petroleum code is a good example of the kind of change Africa needs. The country's passage of the Hydrocarbon Law in 2019 was not just a regulatory milestone, but also a major step forward for the government's efforts to revitalize the oil and gas sector. The new code sets more attractive fiscal terms to investors, lowering royalties and taxes and offering greater flexibility. For the first time ever in Gabon, it also makes special provisions for gas projects.

The Hydrocarbon Law is also breaking new ground with respect to local content. It prioritizes the local workforce in hiring and training. It requires investors to commit to a course of progressively replacing expatriate staffers with Gabonese employees, and it only allows the hiring of foreign workers in circumstances requiring exceptional skills.

Additionally, the law identifies indigenous companies as organizations in which at least 60% of the capital is held by Gabonese nationals and at least 80% of the workforce comprises Gabonese nationals. It then stipulates that companies fitting the description will be first in line for contracts related to the development of both marginal and mature oil and gas fields. And it does this so that local businesses can build up a strong value chain.

And Gabon is getting results: The country signed at least a dozen new production sharing contracts (PSC) with international investors including Sinopec, Petronas, and Perenco in the year after it passed the Hydrocarbon Law. It also struck a USD24 million deal with Assala Energy on the modernization of oil export infrastructure at the port of Gamba in October 2020—and since 30% of the country's crude oil exports pass through Gamba, this project will be critical to the economy.[6] What's more, VAALCO Energy extended its Etame Field PSC in 2018 for 10 more years, allowing for more investment in that project.

The new Hydrocarbon Law has modernized Gabon's fiscal regime and increased its ability to stay competitive. It has also allowed the country to start new ventures without risking existing investments.

Gabon's experience offers an inspiring example of the kind of reform that will drive the growth of the African energy sector and facilitate the energy transition. We need more commitment and strategic action like this from African governments.

Strategic Fiscal Policies: The Foundation of Real Reform

I strongly urge African governments to consider IOC's pain points and eliminate, or at least reduce, the challenges of doing business in Africa. Here are a few measures African governments should be taking:

Rethink licensing rounds: There was already a need for licensing process improvements well before the pandemic, and streamlining is even more urgent in the current environment. While the existing process has important benefits in ensuring companies have sufficient resources for exploration—and protecting their rights—it's also fraught with delays and red tape, creating hardships for companies forced to wait years for decisions. They can't afford to put capex resources aside for that long in this environment without a guarantee.

We can't ask companies to wait that long, or they'll go elsewhere—and they have. There have been several delays and mishaps with licensing rounds that attracted very few or low bids or fizzled out altogether. Algeria and Nigeria have both suffered a loss of interest from investors and declining oil production in recent years because of slow movement on this front. Gabon and Somalia both had to extend 2020 licensing rounds, in part due to delays in enacting key legislation.

I've advocated moving to direct negotiations as a way to offer more flexibility to exploration and production (E&P) companies. Negotiation would not be tied to rigid time schedules the way licensing rounds are, and it would also allow discussion and possible adjustment of contract terms.

Incentivize natural gas exploration: We need to revise our PSCs to give more favorable terms to our investors. Specifically, we need to incentivize gas projects as well as oil exploration and production. Most African countries currently have oil-based regulations or production contracts that don't also include provisions for gas.

A lot of countries have difficulty working with companies to get to FID on natural gas discoveries. We need to develop contracts that define the terms for gas drilling upfront. This will ensure that companies who drill for oil and find gas instead don't have to then go back to the table to renegotiate. We should have natural gas-specific PSCs to make it easier to conduct profitable gas exploration.

The lack of a clear gas drilling policy creates tremendous uncertainty for our partners—on top of the uncertainty that's already inherent in drilling for oil. Why add another layer of confusion and delay in an environment that's already challenging?

Let's provide predictability and consistency. Let's give them every reason to do business with us, regardless of whether they find oil or gas.

Negotiating from Strength: Better Deal-Making

Another crucial building block of governance reform—and where we still need improvement—is deal-making. In *Billions at Play*, I wrote about the fact that we have often failed to get deal-making right—that is, in a way that utilizes our resources to benefit Africans. We need to negotiate better deals if we're going to get out from under the resource curse and realize our potential.

The negotiating advice I outlined in *Billions at Play* bears repeating. It's even more crucial now in the post-pandemic environment:

- Choose the right investors—those with the resources and commitment to actually begin and sustain operations in Africa.
- Invest the necessary time and effort to come to the table prepared with the best understanding of our investors' interests and concerns.
- Keep all key stakeholders and decision-makers involved throughout the process to ensure there are no surprises.
- Make sure the terms are realistic—that the deal can actually be executed when the negotiations are over.
- Be realistic about the complications our investors face in dealing with us. Burdensome tax policies, excessive red tape,

unrealistic local content requirements, lack of judiciary protection for contracts, and lack of transparency are only a few of the issues.

We need to consider the perspectives of both parties so that we strike the strongest possible balance between a fair return for investors and benefits to Africa. Yes, we need to be flexible. But we also need to be pragmatic about our own interests.

I also recommend making some changes in line with the realities we now find ourselves in. One immediate priority is adjusting contracts to meet changing post-COVID demands on the industry and the needs of our investors. We should be actively engaging in contract renegotiation with a focus on these key considerations:

- Both parties' ability to fulfill capex commitments in joint venture or production sharing contracts.
- Flexibility to accommodate capital spending commitments and repayment obligations.
- Crude oil price fluctuation affecting future receivables.
- Ability to get financing.
- Changes in law.
- Political and economic stability.
- Local content obligations.
- Opportunity costs of the deal.
- Flexibility in independent contractor and service provider terms to allow for potential defaults during the post-pandemic period.

It's in our best interest to avoid a situation in which force majeure becomes our investors' only option. At the same time, we should be mindful of placing severe contract limitations on the ability to declare force majeure, since that in itself is a deterrent to investment in this environment.

And, finally, African producer states need to learn to ask for more. They need to lobby for knowledge transfers, training, gas monetization programs, and other significant opportunities. This is how we can strategically manage our oil and gas operations to create pathways for economic growth and diversification.

We have success stories to look to.

Mozambique, for example, is an outstanding model of strong energy sector negotiation designed to benefit the entire region. They've had a rise in 2020 deal activity despite the industry decline. *Africa Oil & Power* named President Filipe Jacinto Nyusi Person of the Year in March of 2021 for his exemplary role in expanding industry and attracting foreign investment, including several multimillion-dollar natural gas projects.

Under Nyusi's leadership, Mozambique has emerged as a regional center of energy investment and development, with a combined investment of more than USD50 billion from TotalEnergies, ExxonMobil, and ENI. More than 68 new U.S. companies signed USD1.8 billion in engineering, procurement, and construction contracts to supply and develop the landmark LNG-gas processing facility.[7]

The growth is expected to continue in 2022 due to favorable contractual terms that include priority for local entities, a public bidding process, and limited government involvement. Service contracts are overseen directly by the National Petroleum Institute and include, for example, the requirement that foreign entities providing goods or services to petroleum operations must do so in partnership with national entities.

Ending Corruption, Improving Transparency

Beyond the regulatory and fiscal reform I've talked about, governments in Africa need to show greater commitment to improving transparency in revenue management and fighting corruption. This remains a major deterrent to potential oil and gas investors for obvious reasons.

Transparency.org's *Corruption Perception Index* (CPI) is an annual report that ranks 180 countries and territories by their perceived levels of public sector corruption. While not specific to the energy sector, the 2021 report painted a bleak picture of Sub-Saharan Africa and described a poor overall performance with improvements made by some countries in the region overshadowed by continuing corruption in others.[8]

The Extractive Industries Transparency Initiative (EITI) has been the single most successful anti-corruption program in the oil, gas, and mining industries. It was founded on the simple premise that companies must voluntarily disclose what they pay to governments from their oil, gas, and mining activities; governments must disclose what they receive; and disparities must be investigated and reconciled.

Since it was implemented in 2003, the EITI has allowed governments in developing countries to recover billions in unpaid taxes, royalties, and other duties. Fifty-three countries worldwide are now members, but it has had a particularly strong impact in Sub-Saharan Africa. The EITI has given voice and recourse to those who are usually left out when the deals are made. The ability to question policymakers is even more important now, given how fast the industry is changing in Africa.

When the EITI announced in October 2020 that USD3 billion had been recovered in unpaid taxes from oil and gas companies to the Nigerian government, it was a stunning example of the impact this initiative can have in improving transparency.

When the administration of U.S. President Donald Trump withdrew the U.S. from the EITI in 2017, it wasn't surprising, but it dealt a blow to the fight against corruption in Africa. Essentially, the message was that the U.S. doesn't care what its companies do elsewhere as long as the interests of U.S. corporations are served. The U.S. government is unwilling to uphold the standards it demands of other countries. This sends the wrong message to the rest of the world. Without the moral and ethical leadership of the U.S. on this issue in support of our efforts, our fight for transparency is an uphill battle.

Currently, 24 African countries are members of the EITI. In 2021, Angola announced its intention to join, and is currently implementing the measures required for membership.

Certainly, any progress in improving transparency is laudable. But more African countries must take decisive and aggressive government action to reform their revenue management. Continued failure to do so will prevent us from ever realizing the potential of the energy transition and our resources.

Taking a Stronger Stand on Human Rights

Widespread human rights abuses across Africa are unfortunately nothing new, and the pandemic further intensified the problem. In an already tumultuous year following elections across the continent, conflict escalated in several countries. With the continuing risk of terrorist attacks, mass abductions, displacement, and countless atrocities taking place across the continent, the crisis is reaching existential proportions and creating an environment of risk, fear, and instability that is a major deterrent to foreign investors and IOCs.

An incident of police brutality by the Special Anti-Robbery Squad (SARS) in Nigeria in 2020 is only one of the more recent instances of the problem. But this incident is significant because of the huge social media backlash and the demand for government action that followed. Thanks to the internet, human rights abuses in Africa now play out on a global stage. And in the current environment of heightened demand for political and social justice, we're being held to a new standard of accountability—not just by potential investors, but by the world at large.

Maybe this will force us to move more decisively to end these atrocities.

Part of solving this problem means embracing the reality of our poor handling of human rights. We can't pretend it's an issue from the past. We can't pretend it's an American or European idea. We are living it, and it's harming our people, our industry, and our economy.

Our governments must address the root causes of the widespread political instability underlying our human rights crisis. Our mandate is to ensure much greater protection, justice, and accountability. Without human rights reforms, the high risk of doing business in Africa will simply send international energy companies elsewhere. We cannot afford for this to happen, as a continent or as an industry.

Ending Resource Nationalism

Too many African politicians are quick to say, "It's our oil. It's our country." They take an "us versus them" attitude, as if "they" are coming in to steal our resources. We need to move away from the "my country, my oil" mindset. It's wrong, and it's hurting our industry and average people.

We must be honest with ourselves and recognize that this resource nationalism—a country's efforts to maintain tight-fisted control of its resources—is just as bad as racism or sexism. Xenophobic feelings about expatriates or foreigners coming into the country to work infect the entire system. And that, in turn, affects the African people we want to protect and empower.

Africans have gone into different waters, been to different places around the world, and we have marched and protested to be treated with decency and respect as human beings. When we operate from a mindset of resource nationalism, we bring the very attitude we have fought against to our own business dealings. We are less likely to create an enabling environment for investment, less likely to be transparent, less likely to develop effective local content policies, and less likely to benefit from knowledge-sharing, capacity building, and valuable public-private partnerships We have never seen a country grow stronger and more prosperous as a result of resource nationalism—and if we wade into that territory, we stand to lose.

Resource nationalism is one reason it took Uganda so long to reach an agreement on Lake Albert oil exploration. It's also the reason that Equinor and Shell LNG have a USD30 billion project in Tanzania that's been suspended since 2019, when former President John Magufuli ordered a review of the PSA regime, specifically clauses related to repatriation of funds, arbitration issues, revenue sharing, and parliamentary power. The Tanzania project seems to be moving forward again under his successor Samia Suluhu Hassan. But it's been very difficult to get anything done there, and part of the reason is this kind of nationalist thinking.

In this day and age, we can't take an adversarial attitude toward partners who want to walk this path with us. We can't make negotiations so difficult. We can't complicate the business environment by only looking at what we can keep for ourselves.

Resource nationalism also led Cameroon and Equatorial Guinea to miss out on a valuable opportunity. Gabriel Mbaga Obiang Lima and his Cameroonian counterpart signed an MOU in 2017 to jointly develop the Yoyo/Yolanda discoveries in a contiguous reservoir offshore Cameroon. Noble Energy, which was just acquired by Chevron, would be the operator. But now they can't develop those resources because the two countries are fighting over who owns what. They need to focus on the economics of the discoveries, get on with these projects, share the spoils, and create opportunities for average people looking to better their lives. I could state the same for the Etinde discovery with Perenco taking over from New Age.

Another example? The maritime border dispute between Kenya and Somalia. They've been locked in a long-running dispute over ownership of a contested offshore area of the Indian Ocean. Of course, this is part of the larger political conflict between the two countries, but the inability to compromise on a solution curbs exploration and delays investment, denying both countries the benefits of the oil resources. So, again, Africa suffers.

We must welcome investors and negotiate with them—and each other—from a partnership mindset with the big picture in mind: What resources can we share? How can we fast-track development? What's the best deal we can make for both of us?

An Urgent Call to Action on Governance Reform

Africa has everything we need to meet the demands of the energy transition and shape our future. I believe in the potential of our industry and the benefits we stand to gain by managing our oil and gas resources strategically. But our governments must take bold, immediate action. We need the kinds of governance reform I've talked about here to accomplish this. We need to address the specific issues I've outlined. We need strong, decisive leadership to institute the right regulatory, tax, and fiscal policies and structural reforms to attract investment and build socioeconomic prosperity from oil and gas. And we must protect our citizens from the negative impacts of poorly managed oil and gas operations—corruption, human rights violations, and revenue mismanagement.

There's greater urgency than ever to make these changes. It's more than an opportunity—it's a profound responsibility. It's up to us to create an energy transition that is truly Africa first.

A tall order, perhaps, but one we can achieve.

Chapter 11

OUR GOVERNMENTS MUST LAY A PATH FOR A SUCCESSFUL ENERGY TRANSITION

FOR THE RESIDENTS of Yaoundé IV,[1] one of the seven communes of the Cameroonian capital, renewables are already a part of daily life. Hydropower makes up an impressive 73% of the power generation mix in the city of Yaoundé IV.

The government of Cameroon has established nationwide plans to reduce carbon emissions by 32% by 2035 with widespread deployment of renewables, particularly via hydropower and solar photovoltaic (PV—rooftop systems). In 2020, its short-term energy and climate action plan vowed to cut carbon emissions and boost energy access by distributing 3,600 solar kits to poor households, installing 3,000 solar streetlights in its 65 neighborhoods, retrofitting 30 municipal buildings with distributed rooftop solar PV equipment, incentivizing the increased adoption of hydro-powered electric motorcycles, and building nine micro-biogas plants.

Further west, Cocody,[2] Côte d'Ivoire, has set a 2030 target to cut carbon emissions by 70%, minimize local greenhouse gas emissions by 90%, and attain the national goal of 42% renewable electricity generation. To do this, Cocody city planners introduced the Green City Plan back in 2017, which involves progressive steps like installing 5,000 solar lamp posts and 1,600 solar traffic lights; equipping 200,000 households with solar PV power kits and 300,000 homes with efficient cookstoves powered by ethanol distilled from local sugarcane; erecting four wind farms on the shore of Ébrié Lagoon; replanting two million mangrove trees; and developing green spaces through reforestation and carbon sequestration. City planners are hopeful that these ambitious measures will simultaneously create hundreds of thousands of new jobs and empower women to participate in the industry.

Meanwhile, citizens of the Ugandan capital of Kampala[3] are increasingly commuting on electric motorcycles that are charged primarily through hydropower. Since introducing its Kampala Climate Change Action Strategy in 2016, the city established an air-quality monitoring and assessment system. A big target of the strategy is the fleets of gas-powered motorcycles used for both commercial and private transport, as they contribute to air and noise pollution and drive gasoline demand. The city has fostered public-private partnerships to foster and fund start-ups that aim to replace conventional motorcycles with electric versions. As of 2020, more than 200 new and retrofitted electric motorcycles have taken to the city streets, silently conveying passengers with scant emissions and requiring almost no infrastructure.

These are just a few examples of cities taking action to combat climate change. Cities and even some nations have also established aggressive plans toward adopting clean energy. This is encouraging, but we need to see even more strategic planning to guide our municipalities, provinces, and nations toward a successful energy transition. And as we're developing these plans, we need to ask ourselves: *Are these strategies in the best interests of average Africans? Will they curtail economic growth or promote it? Will they result in job creation? Will they create opportunities for African businesses? Will they adequately address Africa's widespread energy poverty?*

My point is that African governments must plan carefully and strategically if we want a just energy transition that does more good than harm for Africans. As we gradually move away from fossil fuels, we must ensure that we Africans aren't stripped of the benefits we would see from our abundance of oil and gas, like job creation and economic diversification. We can—and must—take advantage of this transition to benefit the continent and our people.

In short, this must be an Africa-first energy transition.

At this moment, we have a golden opportunity. The pressing discussion of the transition to renewable energy sources in the battle against climate change can be beneficial to Africa and its people—but only if we, as Africans, shape and guide it. We must direct the storyline to make the plot work for us. In the past, we've simply responded to what others say we need. Enough playing defense! It's time we play offense—Africans need to define the *African* agenda on climate change and renewable energy, in a unified *African* voice that demands to be heard.

African governments—along with African businesses—must take the lead. They need to speak up for African interests. And they need to start working now to set the stage for our Africa-first transition to renewables to succeed. But it will take all of us—governments and citizens together, private and public entities alike—to make it work.

Planning to Ensure Economic Growth

Let's return to the idea of planning. Right now, African nations are under immense pressure from the international community to pivot from fossil fuels to renewable energy sources as soon as possible. Appeasing those demands without a solid strategy is a recipe for disaster. African countries must take the time to craft transition plans customized for their particular needs, challenges, strengths, and opportunities.

One of the many factors governments should explore as they develop their strategies is to promote economic growth. African countries must have mechanisms in place to ensure that their economies prosper as the continent sees more green energy activity. And governments must be deliberate in ensuring transparency and good management of renewable energy project revenue.

Some African countries have already made significant strides in these areas. Kenya's Green Economy Strategy and Implementation Plan,[4] for example, ambitiously calls for "achieving an annual growth of 10% per annum and transforming Kenya into a globally competitive and prosperous country with a high quality of life by 2030."[5] The plan also calls for an institutional framework for the coordination and management of green finance flows through coding, tracking, and reporting. These efforts, already underway, are fostering transparency as Kenya's government receives and spends revenue from green energy projects.

According to Kenya's implementation report, "This has resulted in greater confidence from our partners, enabling us to tap into diverse green funding sources such as the Green Climate Fund, the Green Climate Investment Fund, and Green Adaptation Fund, among others."

Kenya has also created collaboration and partnership mechanisms with the goal of encouraging green financing from financial institutions, organizations, businesses, and others. And the country is encouraging ongoing cooperation, idea-sharing, and partnership opportunities as a founder-member of global and African green finance initiations, coalitions, networks, and partnerships, including the Coalition for Finance Ministers for Climate Action and the International Platform on Sustainable Finance.

Will Kenya succeed? I don't know. But *without* economic goals and a roadmap for reaching them, African countries most certainly will experience even more economic hardship during the energy transition.

Technology Transfers and Skills Development Are Vital

If we want to fully reap the economic benefits of expanded green energy operations, we must take an honest look at our current limitations, specifically, the gaps in our green energy technologies and skills. Without dramatic improvement in these areas, Africans will lag behind foreign companies and employees who are all too willing to capitalize on green energy opportunities on our continent.

The recent International Labour Organization report, *Skills for Green Jobs in Ghana*,[6] addresses this point. "Currently there is a significant skills gap, particularly among high skilled-semi-skilled and vocational/technical skills and this has implications for hampering the green economy agenda. Training in science, technology, engineering, and math (STEM), which is critical for speedy transition into green activities in energy, forestry, agriculture, manufacturing and so forth, is quite limited," the report stated. "It is recommended that a comprehensive and cohesive approach to developing green skills is adopted at all levels of the education system through a review of the current education and skills training framework in line with the country's greening aspirations."

Ghana is hardly alone in this predicament. In a 2020 article, the Strathmore University Energy Research Centre in Kenya noted that technology transfer will be necessary for a domestic workforce with a full spectrum of green economy skills. "One way of looking at technology transfer is it makes it possible to have people in the country with the skills to be able to work on equipment, technology, and plants that support the green economy and that they can install, use, maintain, repair and re-cycle," the article stated.

The report continues by stating that, "sadly, this is not the case."[7] Kenya has not kept up with the pace of technology transfer. Instead, during the last 10 years, Kenyan renewable energy companies have imported equipment, flown in experts to train local people to use the equipment, and then assumed that technology transfer had occurred. But as the equipment broke down, experts had to be flown back to Kenya for repairs. "This leads our engineers to become users or installers because they cannot repair or re-use or produce," the article said. "This is the gap."

Ameenah Gurib-Fakim, former president of Mauritius, and Landry Signé, senior fellow in the Global Economy and Development Program and the Africa Growth Initiative at the Brookings Institution, put it in even bleaker terms, noting that Africa contributes only 2% of world research output, 1.3% of research spending, and a dismal 1% of all patents. In a report they co-authored, Gurib-Fakim and Signé state: "A lack of investment in science and technology has undermined Africa's economic transformation at both the structural level (the shift of workers and resources from low- to higher-productivity sectors) and the sectoral level (the growth of productivity within sectors)... This lack of investment has had far-reaching consequences: Without the economic and scientific infrastructure necessary for innovation, the continent has continued to rely on the colonial development model of resource extraction, which is both unsustainable and largely responsible for its debilitating poverty and aid dependency."[8]

How can African governments change course? They must commit to policies that foster STEM education and develop more green energy vocational programs. They must encourage African innovation and technological development, particularly in areas of renewable energy. They must encourage private sector innovation and public-private-academic partnerships that span borders. African governments absolutely must encourage international oil companies to invest in African projects, especially natural gas, which provides critical opportunities for technology transfers.

Addressing this gap must be a priority for African governments, one that they tackle in cooperation with businesses, educational institutions, and other organizations.

African Owned and African Operated

As we work together to protect the world from climate change, we must be pragmatic about life on the ground for Africans. If we embrace an energy transition that creates lucrative opportunities for non-Africans and leaves the vast majority of our people and businesses on the sidelines, we will have failed. It would be disastrous to allow the economic opportunities oil and gas represent to wither on the vine, without putting new opportunities in their place.

Today, everything is produced in the West and sold to Africa—why not switch roles tomorrow?

Along with budgets for training, governments must have strong local policies in place to ensure jobs for African workers, opportunities for African companies to provide products and services for renewable energy operations, and knowledge transfers so that more African companies and entrepreneurs can be competitive in this sector.

Governments must actively work to bring the manufacturing of renewable technologies to the continent. Today, everything is produced in the West and sold to Africa—why not vice versa tomorrow? Governments must ensure that Africans get the lion's share of jobs created by the energy transition—through the entire renewables value chain.

This is the critical caveat: Job creation must not be limited to entry-level, low-paying positions. Africans must have the opportunity for solid careers in the energy industry and adjacent sectors. They must have access to successful pathways to promotion, eventually leading to leadership positions. In short, Africans must be the ones building the factories as well as learning and implementing the technologies, designing the new factories, managing the development, and running the energy companies.

Governments must make sure that African companies are building Africa's new renewable energy infrastructure, facilities, and equipment. To do this, each nation needs strong, enforced local content requirements. Each nation needs to establish a viable energy transition plan with a framework to empower indigenous companies to fully capitalize on renewable energy industry opportunities.

The way forward is to set up the energy sector for greatness by establishing strong precedents while renewables are still in their infancy. We will stimulate economic development, attract investments, and create jobs for young Africans.

In this regard, Africa has an advantage on the world stage: a young population. If we give our young people the power, they will make history.

Admittedly, we currently lack the skills and resources needed to act immediately. We need to change that post-haste! We need to develop and nurture future generations through education and training programs, knowledge-sharing with the international community, innovative programs by non-government organizations (NGOs), and other creative solutions.

We won't succeed unless our people have the training they need.

Let's look to Open Africa Power,[9] for example, an educational initiative led by the Enel Foundation in cooperation with several prestigious universities throughout Africa. This high-level training program offers knowledge on all aspects of sustainable electricity production and distribution for master's and doctoral students in engineering, law, and economics. A key element is that the program also facilitates networking for participants to continue sharing knowledge after the formal classroom segments are complete.

In 2018, the inaugural Open Africa Power program launched at Strathmore University in Kenya with 29 participants in the six-month training program that combined classroom lessons and e-learning with onsite business visits. The program continues to have a positive impact.

According to the Enel Foundation, "Open Africa Power has seen a constant and remarkable increase in the number of applicants, from 140 in 2018 (its first year), to over 750 in 2020; almost 40 African nations were represented. Last year's 61 selected candidates were highly skilled (63% of them held, or were in the process of obtaining, a PhD), and showed diverse and complementary backgrounds, ranging from IT to law, from engineering to economics."[10]

After attending the 2022 program, Sokoyebom Alabi, a power utility engineer in Nigeria, said, "I have always thought that Renewable Energy was the answer to sustainable energy access. However, I have learnt something different. Carlo Ferrara opened my eyes to the fact that renewable does not mean sustainable. From Georgia Scaffardi, I learnt about Sustainable Power plant and from Quadri Suelen that an Independent Power Plant (IPP) can and should create economic development throughout its lifetime."[11]

Our young professionals are clamoring for this knowledge. We must ensure that we continue to deliver opportunities for youth to attain it. Let's be sure that Open Africa Power isn't the only vehicle delivering this vital training in the continent.

African NOCs Must Have a Voice in the Energy Transition

Africa's national oil companies (NOCs) must have a voice in their countries' energy transitions, too. These state oil companies will need to adapt to a downward progression in revenue following decreases in fossil fuel consumption, and they need to have a stake in the opportunities created by the renewables market. Some NOCs have already introduced reforms to diversify their portfolios, increase efficiency, and reduce exposure to market decline. But as renewables become the dominant energy sources, all NOCs need to play a growing role in decisions about how carbon abatement technologies are developed, introduced, and required.

Given that their main role "is to monetize national hydrocarbon resources as effectively as possible in order to maximize state revenues while guaranteeing a secure supply of affordable energy for the nation's needs," Raad Alkadiri and Björn Ewers wrote in a piece for Boston Consulting Group, "the revenues produced by those organizations are vital for economic sustainability. They help to finance public infrastructure projects and development and welfare programs. The NOCs themselves play a key role in developing national resources and capabilities and in supporting employment."[12] This means that NOCs can't simply be cast aside if and when petroleum is no longer in vogue—they will remain a critical component in the resiliency of their economies in the wake of the energy transition.

What if every African petroleum ministry were charged with coming up with answers to the following questions: *What is your strategy for low-carbon energy production? What is your strategy for the energy transition? And, most importantly, how are you working on it with your citizens?*

The oil industry has the technology and the capital to turn national *oil* companies into national *energy* companies. We must put the onus on them to do things differently, and better. We must push them to operate their assets more efficiently, to help decarbonize energy and fast-track a move to gas while also growing their business within Africa. We must encourage them to invest in developing carbon capture and storage technology within Africa.

Policy matters: We cannot afford to make the mistakes of the past.

Please Tell Me We've Learned from Our Mistakes!

When Africa's vast petroleum reserves were first discovered, then extracted, the fledgling energy industry wasn't sufficiently regulated. Our governments failed to establish strategies for oil and gas production or to safeguard the interaction with and involvement of foreign investors. Amid the excitement of oil and gas discoveries and their potential to generate revenue, inadequate regulatory frameworks did little to protect African economies, communities, or businesses.

While some countries have created local content legislation in attempt to ensure jobs for Africans and opportunities for African businesses, we have seen an uneven effort across the continent since those early days. Meanwhile, we've seen widespread red tape and corruption. As a result, we've also seen the wealth from under our soil pulled out from under our feet for the profit of warlords and foreign entities.

Now that renewable energy is on the horizon, we have the opportunity to hit "reset." We must ensure that we act smarter this time to avoid another "resource curse." We CANNOT go down that path again. We can't succumb to poor revenue management that results in a failure to monetize. And we can't afford to let energy revenues line the pockets of a privileged few outsiders while millions of Africans fail to realize their benefits.

I'll discuss electric vehicles (EVs) elsewhere in this book, but I feel I cannot discuss good governance related to renewables without briefly mentioning them here. Projections by the International Energy Agency (IEA) see yearly EV sales topping 44 million vehicles before 2030, resulting in skyrocketing demand for the rare earth minerals that help power these vehicles.

For better or worse, Africa holds a wealth of mineral resources needed for EV batteries,[13] including:

- Zimbabwe: vast lithium deposits.
- Namibia: lithium.
- Mozambique: some of the world's largest stores of graphite.
- Tanzania: graphite.
- South Africa: the largest known reserves of manganese and platinum in the world, along with considerable nickel deposits.

- Zambia: copper.
- Democratic Republic of Congo (DRC): copper, lithium, and 73% of the world's cobalt production.

Being resource-rich, we've found, isn't always a pathway to prosperity. In fact, in many African petrostates, resources have been more of a road to ruin. The first time around, African governments were at fault in their complacency. We can't allow them a second shot at mediocre legislation—otherwise, we face a repeat of the rather paradoxical stunted economic growth and rise in authoritarian regimes that have plagued countries with an abundance of highly desirable natural resources much more than countries with fewer natural resources.

To ensure that Africa and its people benefit from the world's energy transition, and specifically the increased EV usage, governments must develop strategies that provide protective regulations for mineral exploitation. Africans need to control the fate of our minerals, with African operations controlling the industry and Africans working—and leading—these operations from extraction through processing.

Without a doubt, mineral exploitation on African soil is going to play a big role in the global energy transition. Africans must be included and involved at every step. We need to be very strategic with these minerals and pay a lot of attention to foreign interests.

Listening to Good Advice and Sharing Knowledge across the Industry

Governments must solicit input from non-government entities with knowledge of the energy industry in Africa—and then put this advice to good use! The private sector is not the enemy; business is our friend.

Lawmakers must engage with energy-sector stakeholders like local network operators, as well as cross-border and pan-African organizations. The African Energy Chamber that I helped establish works to empower stakeholders through education and awareness, but it is hardly the only organization with this objective. The United Nations Economic Commission for Africa, for example, has established a Technology, Climate Change, and Natural Resource Management Division (www.uneca.org/TCND) that conducts research to support lawmakers as they draft logical regulatory frameworks for the proper management of natural resources in Africa. These efforts increase stakeholder participation regarding Africa's management of its mineral resources and its development of renewable energy by fostering dialogue among stakeholders to promote the exchange of experiences and best practices across the industry.

Additionally, the United Nations runs the African Ministerial Conference on the Environment (www.unep.org/regions/africa/african-ministerial-conference-environment) (AMCEN) to facilitate important dialogue on global and regional environmental topics. With an Africa-centric slant, the group fosters the active involvement of Africans in such discussions to keep the continent and its people well-represented in negotiations on international agreements. AMCEN draws on the expertise of African representatives to advocate for local programs, introduce knowledge-sharing and training opportunities, and provide strategy and policy guidance for governments across the continent.

And we can't avoid mention of the African Continental Free Trade Area (AfCFTA), which made history on January 1, 2021, by opening trade among 54 of the 55 African Union member nations.[14] The free trade agreement removes tariffs from 90% of goods, essentially allowing a free flow of goods and services across the continent. The AfCTA creates a single market for stronger economic integration, fosters investment and industrial development through diversification, and enhances the competitiveness of member states in the global market. And this has been a long time coming, as far back as 1963 when the Organization of African Unity was launched with the goal of an African common market.

Breaking down borders is critical! The policy of requiring visas and impeding travel from one African country to another is antiquated. Just look at what the European Union has done to tear down the walls of restrictions and create an oasis of transitions and movement. I sincerely hope that the AfCFTA will emulate this success.

I wholeheartedly second the opinion of Mohamed Béavogui, director general of African Risk Capacity, who stated, "Africa needs to move towards market-based innovative financing models to achieve a strong, united, resilient and globally influential continent. The future of Africa depends on solidarity."[15]

Through solidarity, we Africans can help the world understand that we have a rightful place at the world energy table. Our continent can no longer be overlooked. We will make a difference.

Calling All Citizens of Africa!

We can't just wait for our governments to take care of us. We all have a role to play in ensuring we reap the benefits. We must be more dynamic. We must stop complaining about our problems and assigning blame to others for those problems. Yes, problems do exist. But we can do something about them other than bemoaning all we don't have, all that the West didn't give us.

Admittedly, the deck is stacked against us. All the more reason we can't give up. This is not the time to shrug our shoulders in resignation and then bury our heads in the sand—it's time to pull ourselves up by our bootstraps. We need to be able to say that we determine our own future. Firstly, individual citizens have a personal responsibility to make the transition successful: Take the initiative to get trained for opportunities in renewable energy operations.

Look for clean energy entrepreneurs and support them. Find allies in different places; don't shy away from alliances with Westerners simply because they're not African—that's just wrong. And remember that African governments are beholden to African people. It's up to you to make sure your government is proactively working toward an Africa-first energy transition.

In addition, African companies need to take responsibility for building up our communities. Look to create public-private partnerships with government allies to move forward. Start by applying for concessions from your governments for local renewable energy projects like wind and solar. Winning those grants will get your foot in the door, prove to investors that you are viable, and show your communities that you are dedicated to their success.

Chapter 12

OPEC—YES, OPEC—MAKES SOME GREAT POINTS ABOUT ENERGY AND CLIMATE CONCERNS

IN 2019, MOHAMMAD Sanusi Barkindo, secretary general of OPEC at the time, responded to young protesters around the globe who were demanding immediate action to curb climate change, including reductions in the consumption of fossil fuels. He acknowledged their concerns but described their campaign against oil and gas as unscientific, noting with dismay that environmental activism was beginning to "dictate policies and corporate decisions, including investment in the industry." He called this shift "perhaps the greatest threat to our industry going forward."

His statements drew a gleeful response from Greta Thunberg, the teenage Swedish activist who has participated in and inspired climate-focused school strikes around the world. In a Twitter post that quoted Barkindo, she wrote: "Thank you! Our biggest compliment yet!"

The exchange made headlines, and Thunberg's tweet received tens of thousands of likes.[1] I find it disheartening that so many people, particularly in the West, view the disparagement of the oil industry as a win, and treat pushback from an organization like OPEC as a badge of honor for any self-respecting champion of the environment.

Yes, OPEC clearly has a vested interest in keeping the oil industry healthy. However, the organization is not an enemy of the environment. It is not a climate-change denier, and it is not obstructing climate-related policies. But it is *realistic* about the continuing role of oil and gas during the energy transition in a way that climate activists are not.

Barkindo himself, who died as this book was being finalized, has been speaking about the need to address climate change for years. To cite just one example, consider his statement at the 2019 UN Climate Change Conference[2]: "We recognize the complexity and magnitude of climate change we are living in our countries. There is no panacea for global warming. All viable mitigation and adaptation measures are necessary."

It is important to note, though, that Barkindo referred to "viable" measures. He didn't recommend an all-out sprint or a frenzied dash toward the finish line. Instead, he wisely encouraged a steady, purposeful jog toward the finish line with respect to making the transition to green energy sources. That's because OPEC believes, as do I, in following a strategic path—in changing the energy mix gradually by adding renewables to fossil fuels, rather than abandoning oil and gas prematurely.

Barkindo made his position clear at the 2019 UN Climate Change Conference. "We reject the misleading narrative of an energy transition from one source to another," he said. "The energy transition must be holistic, inclusive, fair, and equitable in accordance with the core UNFCCC (United Nations Framework Convention on Climate Change) principle of common but differentiated responsibilities and respective capabilities."

So instead of demonizing oil and gas, Barkindo and other OPEC leaders are encouraging us to recognize their tremendous value. They are talking about the industry's potential to contribute to economic growth and diversification in oil- and gas-producing nations. They highlight the crucial role that natural gas can play in initiatives to eradicate energy poverty.

What's more, they acknowledge that the oil and gas industry can make significant contributions to climate solutions.

For one thing, oil and gas operators can advance relevant technologies such as carbon capture, utilization, and storage (CCUS), which involves extracting carbon dioxide emissions from power plants and factories and then storing or reusing them. Indeed, they have already done so in places like Brazil, which launched the world's first CCUS-enhanced oil recovery (EOR) project at the Lula offshore field in 2011.[3] Projects such as these are important, given that removing carbon dioxide from the atmosphere is considered a necessary component of achieving the Paris Agreement's climate standards, and they shouldn't be ruled out just because they're coming from oil and gas operators.

For another thing, the oil and gas industry also has an important role to play in generating the revenue necessary for fossil fuel-producing nations to finance renewable energy initiatives and meet global emissions-reduction goals. After all, this transition isn't going to be cheap, and the world must cover the cost. So let's not rule out oil and gas sales as sources of funding just because of their origins. (Really, is wind energy any less renewable when the turbines are purchased with revenue from oil exports?)

OPEC is making valid points. It's calling for a balanced approach, one that recognizes the seriousness of climate change and the importance of addressing it without neglecting the very real needs of oil- and gas-producing nations. Its message is particularly relevant to Africa's petroleum-producing countries, which still have much to gain from the strategic development of their oil and gas resources.

So rather than dismissing the insights OPEC offers because it comprises oil-producing countries, I suggest listening to what the organization has to say.

An Evolving Role in Climate Protection

When OPEC was formed in 1960, its primary goal was to help members make the most of their petroleum resources. And certainly, profit and protecting market share are key OPEC objectives. Since OPEC hasn't strayed from these goals over the past decades, it doesn't exactly come across as a group that can be expected to embrace environmental protection.

Accordingly, it's drawn plenty of criticism for its attitude toward hydrocarbons. For example, when it participated in G77 (a United Nations lobbying group comprising developing nations) climate negotiations with the UNFCC in the 1990s, some of its member countries were accused of engaging in delaying tactics.

One of those leading the charge was Suraje Dessai, a professor of climate change adaptation with the Sustainability Research Institute at the University of Leeds. He addressed OPEC's actions in a 2004 report for the World Wildlife Fund (WWF) titled *A Political Analysis of the Role of OPEC as a G77 Member at the UNFCCC*.

"OPEC and in particular Saudi Arabia have close associations with the oil industry, in particular US companies," Dessai wrote. "This has led them to oppose greenhouse gas reductions, disrupt the whole negotiation process (by asking for equal progress on all issues) or by holding certain issues that are important to other G77 countries (e.g., adaptation) hostage by linking progress on them to progress on the impacts of response measures. This has created some resentment and frustration on the part of G77 delegates."[4]

Today's OPEC doesn't deserve such harsh criticism. OPEC understands that the desire to prevent climate change and appreciate what oil and gas can achieve do not have to be mutually exclusive. Moreover, Barkindo himself understands the issues at play, as he helped produce the UNFCCC and the Kyoto Protocol (a treaty requiring participants to reduce greenhouse gases) as the leader of Nigeria's technical delegation to the UN negotiations in the 1990s. During the G77-UNFCC climate negotiations, he served as the G77 and China chair. This sounds like someone who valued, and did not obstruct, the negotiation process.

I've also noticed that, in recent years, OPEC has been gradually building bridges with the environmental community, and its efforts seem to have been at least a little successful. During the late 2010s, for example, UN Climate Change Executive Secretary Patricia Espinosa spoke favorably about OPEC and said she, too, was in favor of a gradual energy transition. "We recognize the central role the oil and gas industry has played—and continues to play—in the lives of people everywhere," Espinosa said at OPEC's 2018 International Seminar. "It has fueled our greatest achievements and helped us surpass our greatest challenges. It has created jobs for millions throughout the world and raised its standard of living. It has helped build our cities, communities and infrastructure."

Even as she expounded on the importance of the energy transition, Espinosa never suggested that the end was near for the oil and gas industry. "While this transformative shift represents enormous opportunity, let me be clear: Nobody is suggesting that this transformation must be done overnight," she said. "We recognize fossil fuels will remain part of the energy mix for most nations in the foreseeable future. Instead, this transformation to a new, more diverse energy mix must be a progressive one."[5]

Well said! Espinosa's balanced perspective is encouraging, and so is the potential for continued cooperation between OPEC and UN Climate Change, the United Nations entity responsible for supporting the global response to climate change. I hope we see more of it.

Meanwhile, OPEC's support for climate protection has gone beyond public statements. In its first legal workshop for energy industry members in 2019, the organization shared best practices for, among other things, navigating the energy transition. And in 2020, it held a technical workshop on climate response measures that drew speakers and participants from both OPEC and non-OPEC oil-producing countries.[6]

Demonstrating Climate Commitment Among OPEC Members

Meanwhile, even as OPEC takes measures to encourage a successful energy transition, many of its individual member states are already working independently to expand their energy mix. For example, the United Arab Emirates (UAE), which is OPEC's third-largest oil producer, has ambitious plans for a green future. It has been investing in wind and solar developments,[7] opened the Arab world's first nuclear power plant in 2020,[8] and announced plans to become a major producer of green and blue hydrogen.[9] (You can read more about this in the hydrogen chapter of this book.)

African OPEC members are adding renewables to their energy mix as well. There are solar energy projects in Angola, Nigeria, and Gabon, and the Sendje hydroelectric power plant is under construction in Equatorial Guinea. (When finished, the latter facility will be able to supply energy to all of the towns on the country's mainland.)[10] Algeria, meanwhile, has set a goal of bringing its renewable generation capacity up to at least 22 gigawatts by 2030.[11]

Additionally, OPEC member companies have been actively pursuing CCUS projects. Saudi Arabia launched the Uthmaniyah demonstration project in 2015, saying it hoped to capture 800,000 tons of CO_2 per year at the Hawiyah natural gas processing plant and transport it 85 kilometers (52.8 miles) by pipeline to the depleted Uthmaniyah area of the Ghawar field. Saudi Aramco, which is running the project, has called it "one of the Middle East's largest of CO_2 capture and storage demonstration projects." In Africa, Algeria delved into carbon-capture technology in 2004 with its In Salah project, and Angola is exploring proposals for utilizing its deep aquifers as carbon-capture facilities.[12]

These examples and others represent promising steps forward toward climate protection. But as Barkindo has said, the path should not involve abandoning fossil fuel production.

For one thing, Africa will need every solution available to address energy poverty.

"Nobody Should Be Left Behind"

My chapter on energy poverty makes it clear why I'm convinced we must harness Africa's natural gas resources to help alleviate it. OPEC's leadership seems to agree, as it has taken the position that we cannot ignore the needs of millions in our zeal to prevent climate change.

I think that's worth saying again: *We cannot ignore the needs of millions in our zeal to prevent climate change.*

Barkindo stressed this point in 2020 during a speech at the International Carbon Capture Utilization and Storage Conference. "We must be inclusive and equitable in our decisions, and not forget our fellow citizens who, day in and day out, struggle for access to energy the rest of the world takes for granted," he declared. "The facts are sobering: There are more than 800 million people around the world who lack access to electricity and almost three billion who live without modern fuels for clean cooking. Let me be clear: Nobody should be left behind in the energy transition."

The secretary general revisited the topic in January 2021. While addressing the S&P Global Platts Americas Petroleum and Energy Conference, he noted that failing to rectify energy poverty not only hurts those without reliable power but could also lead to unrest. "If billions of people in the developing world suffering from a lack of energy access feel they are excluded from access to energies that have helped fuel the developed world, then this could sow further divisions and expand the divide between the haves and have-nots, the global North and the South," he commented.

But OPEC isn't just raising awareness of energy poverty. It's been battling it through the OPEC Fund for International Development (OFID), which works to stimulate economic growth and social progress in low- and middle-income countries around the world.

One place where OFID has made an impact in Africa is Côte d'Ivoire, where it's funding the construction of the 390-megawatt Atinkou power plant. As of 2017, only 66% of the population of Côte d'Ivoire had access to electricity. The natural gas-fired power plant will use highly efficient combined-cycle turbine technology to reduce generation costs and greenhouse gas emissions, in part, by displacing older generation units.[13] OFID is contributing to a €303 million financing package for the €404 million project. Meanwhile, in Mali, OFID is contributing €85m to cover the cost of developing, constructing, and operating a greenfield 90-MW heavy fuel oil power facility. Less than a quarter of the 17.6 million people living in Mali have electricity, and in rural areas, the share drops to 13%.[14]

OFID doesn't just serve as OPEC's proxy for using oil and gas revenue to reduce energy poverty. It also supports efforts to diversify the world's energy mix. For instance, it is contributing USD18 million to the USD143.7 million Boulanouar windfarm in Mauritania. This project is expected to expand electricity supply in the cities of Nouakchott and Nouadhibou, and it will also help Mauritania meet growing domestic energy demand and produce a surplus that can be exported. Additionally, OFID has made €50 million available to the €1.2 billion Nachtigal Hydro Power Company (NHPC) in Cameroon. Once in operation, NHPC's hydroelectric plant is expected to provide one-third of Cameroon's electricity needs.[15] OFID also supports renewable energy projects in other parts of the world, including Asia and Latin America.

Prophets of Doom, Realities on the Ground, and Multiple Solutions

How fast should the transition to green energy happen? The COVID-19 pandemic has led a number of activists and analysts to raise questions about the prognosis for oil and gas. Some have asked whether the drop in fuel and energy demand that occurred in 2020 will shorten the petroleum industry's lifespan. Others have gone further, declaring that COVID-19 signaled the beginning of the end for fossil fuels and predicting that the energy transition is about to shift into overdrive.

Not so fast, says OPEC.

Barkindo stressed this point during a speech in late January 2021. "It is important that we do not get caught up in the false narratives of the prophets of doom—those that say the oil industry is on its last lap and that peak oil is upon us," he said at the International Carbon Capture Utilization and Storage Conference, held in Riyadh, Saudi Arabia in 2020. Instead, he explained, OPEC is cautiously optimistic that the global economy will rebound this year and that demand for oil will start increasing.[16]

He also said OPEC was hopeful about the oil and gas industry's long-term prospects, even in the face of plans to expand renewable energy generation. He pointed out that the size of the global economy was expected to grow more than twofold between now and 2045, while the world's population is slated to grow by more than 1.7 billion people over the same period.

Growth is likely to support continued demand for oil, he said, and will also increase the need to bring light, heat, power, and cooking fuel to billions of people. "There are obviously many facets to the future energy transition, but the basic challenge is simple: How can we ensure that there is enough energy supply to meet expected future demand growth, and how can this growth be achieved in a sustainable way, balancing the needs of people in relation to their social welfare, the economy, and the environment?" he asked.

Barkindo also pointed out that renewables cannot save the day on their own. And he's right about that: Although renewables are set to leap forward, they still have a long way to go.

That's why OPEC recommends continuing to add renewable energies to the mix while continuing to harness oil and gas to meet the world's needs—and to continue paving the way for the time when renewables can shoulder the full burden.

"We need to understand that the scale and complexity of climate challenge means that no single energy source can solve the problem," Barkindo said. "The fact is, the oil and gas industry—with its extensive know-how and deep technological expertise—is uniquely positioned to be a solutions provider in our common goals of reducing the environmental footprint."

That means, Barkindo added, that OPEC, along with its non-OPEC partners in the OPEC+ group and the entire global oil industry, will have roles to play in providing the solutions the world needs for a sustainable energy future.

I, for one, am glad to hear him say it. The world needs energy, and it needs OPEC—and the petroleum industry at large—to diversify.

Chapter 13

CREATING OPPORTUNITIES FOR WOMEN EMPOWERS ALL AFRICANS

Salma Okonkwo isn't shy about extolling her capabilities, or those of other women in the energy field. "Women know how to move mountains," she recently said. "We're doing so right now."

Okonkwo's own record of mountain-moving is nothing less than remarkable. In the 2000s, she created an energy company to bring liquified petroleum gas (LPG) to Ghana's northern region, the country's poorest area. In less than 20 years, the start-up became a thriving conglomerate named Blue Ridge Group. Its companies include Blue Power Energy, an energy services company that focuses on renewable energy and also operates in the agriculture, consulting, health, petroleum, and security sectors.[1]

Okonkwo, who was named Female Energy Personality of the Year at the 2019 Ghana Energy Awards, credits her achievements to determination. After she struggled to secure funding for her new venture, she refinanced her family's properties to raise the money she needed. Now, her company is making a significant impact on northern Ghana.

Okonkwo is driven by a passion to create employment opportunities, particularly for women, and to bring affordable energy to northern Ghana through Blue Power Energy's solar farm project. "I don't stop when the door is being shut. I find a way to make it work," she told *Forbes*. "That's what propelled my success."[2] I have no doubt that she'll be successful on both fronts.

Not only is she an energy industry mover and shaker, but she's also a much-needed role model for African women. It's no secret that the global energy sector is dominated by men. In 2019, the Petroleum Equipment and Services Association reported that women make up only 15% of the total oil and gas workforce. Even fewer women hold entrepreneurial and leadership roles. And when you do encounter women in professional energy positions, far more work in human resources and communications than in engineering or finance.

Compared to oil and gas, the global picture for women in renewable energy looks like progress—the International Renewable Energy Association (IRENA) reported in 2019 that women held about 32% of the industry's full-time positions[3]—but that only shows how much further women must go to achieve equity in hiring. We need to see considerably more instances of women creating opportunities and driving success in our energy sector.

What's behind the dearth of women in positions in the energy industry? The list is long and complex and includes cultural norms, lack of educational opportunities, regional violence, poverty, and energy sector failures to recruit, retain, and promote women. Some of these are easier to overcome than others, but the fact that women are still underrepresented, underpaid, and under-promoted in energy simply isn't acceptable.

I wrote about the oil and gas gender gap in my last book—but as we expand Africa's energy mix, I feel the need to revisit this topic. We must put a stop to Africa's energy industry gender disparity. We need to create more opportunities for women to join the sector, from oil and gas to renewables, in higher-paying jobs. The effect is good for everyone: women, their families, their communities, and the world.

Positive Impacts

Research proves that when women have access to meaningful employment, more rights for women and gender equality follow. The ability to earn their own incomes and maintain their independence creates personal empowerment with far-ranging effects. For example, the United Nations said, "When more women work, economies grow. Women's economic empowerment boosts productivity, increases economic diversification and income equality in addition to other positive development outcomes."[4]

What's more, women tend to have a positive impact on the businesses they work for and lead, from improved communication and innovation to profitability. After the Peterson Institute for International Economics surveyed 21,980 firms from 91 countries, for example, it found that having women at the C-suite level significantly increased net margins.[5]

Other researchers have reported similar findings. First Round Capital, for example, says that companies with female founders or leaders performed up to 63% better than male-led companies.[6] The CFR Women and Foreign Policy Program's new digital report, "Growing Economies Through Gender Parity," brought these arguments closer to home. The report, which incorporates data from the McKinsey Global Institute, maintains that Nigeria's gross domestic product could grow by 23%—or USD229 billion—by 2025 if women participated in the economy to the same extent as men. And the International Monetary Fund (IMF) has found that strengthening gender equality in Nigeria could be an economic game-changer, leading to higher productivity and greater economic stability.[7]

The benefits of closing the gender gap don't stop there. Because of their experience and backgrounds, women bring different perspectives on a wide range of issues to the workplace. Introducing more diversity of thought into group problem-solving situations, in particular, can resolve complex issues faster and better—improving overall productivity. Imagine how beneficial it would be to have a plethora of possible solutions to test and choose from because there is more diversity on the team.

According to Elizabeth Rogo, the founder and CEO of Tsavo Oilfield Services in Kenya, new perspectives stemming from gender diversification can lead to more innovation and creativity in the energy sector, along with social benefits: The more female leaders in the industry, the more likely women are to consider similar careers themselves.[8]

When we educate women to succeed and hire for positions based on merit without regard to gender, we promote equality and encourage success for all of Africa.

So, as Africa strategically grows its energy mix and we work to create opportunities for Africans, we must make sure that we are doing the heavy lifting to ensure women can seize those opportunities, too.

One major roadblock for women in Africa's workforce is a lack of access to education. Consider the lack of even primary education for girls in Sub-Saharan Africa: Human Rights Watch reported that in nearly 90% of the region, one-third of primary-school-age girls aren't in the classroom. Sadly, the numbers at the secondary school level are equally grim. Two issues related to the region's culture of early marriage keep teenage girls out of school: 40% of Sub-Saharan girls marry before age 18, and then many are expelled because they are pregnant or young mothers. Additional problems include bullying and sexual violence against girls while they are commuting to or attending school, lack of funding for girls' education—it's the lowest budget priority in too many African countries—lack of privacy and sanitation at schools, and cultural norms that place little value on girls and/or girls' education.[9]

And then there's the challenge of giving female students a solid foundation in science, technology, engineering, and math (STEM), which is necessary for most mid-level and higher positions in the energy industry. This is not just an African issue: Globally, women comprise only 35% of all students enrolled in STEM-related fields at the higher education level, according to a 2017 report by the United Nations Educational, Scientific and Cultural Organization (UNESCO). In Sub-Saharan Africa, the numbers are worse, with women comprising just 28% of those pursuing STEM careers.[10]

Gabriele Voigt, president of Women in Nuclear (WiN) Global, which has chapters in Nigeria and South Africa, has suggested introducing girls to STEM subjects at a young age, starting in kindergarten. "Everything starts with education," she said. "One of the most important issues to get girls to study mathematics and physics is to start very early."[11]

Consider the example of the Making Ghanaian Girls Great program, which harnesses solar-powered and satellite-enabled distance learning infrastructure to connect students, teachers, communities, and government officials with interactive learning sessions. The program, which receives funding support from the Department for International Development's Girls' Education Challenge and the Varkey Foundation, has already benefitted more than 36,000 students and is aiming to expand its reach to more than 18,000 primary school, junior high school, and out-of-school girls during the next 4 years.

The effects of programs like Making Ghanaian Girls Great is inspirational. The Varkey Foundation recently reported that participants' fluency in English progresses at a rate of between 30-35% faster than other pupils', while improvements in math scores are 2.5 times better compared to similar interventions. The project also delivers an after-school girls' club called Wonder Women for marginalized girls that is intended to increase participants' self-esteem and encourage them to continue their education.

It's not acceptable that Africa is behind in STEM education. We should be seeing many more STEM-related education programs, from girls' clubs to science competitions to job-shadowing and tutoring programs. I'd also like to see companies and universities working together to attract girls and women to STEM majors, including programs that traditionally draw males, like engineering.[12]

How will we get there? It will require cooperation among governments, educational institutions, and non-government organizations. We need to find ways to provide girls with affordable learning opportunities that incorporate but aren't limited to STEM, as well as safe spaces and important technology.

The Business of Building Equitable Opportunities

Unfortunately, STEM education and role models can only go so far to help women overcome barriers. The corporate world has to do its part as well, and throughout the energy value chain, the effort is seriously lacking. African businesses aren't doing enough to recruit, train, retain, or promote women at any level. Women in the workplace deal with an onslaught of everyday challenges including a lack of childcare resources and maternity leave programs, pay issues, and minimal training and development programs. These issues create systematic problems that contribute to women being underpaid and underemployed.

Part of the problem in Africa is a failure to recognize gender disparity as a problem. A 2016 report by McKinsey noted many African organizations do not take gender issues seriously, with only one in three companies in the private sector citing gender diversity as a CEO priority.

"Today's women leaders have succeeded, it seems, largely through a combination of opportunity and drive rather than through a coordinated corporate effort to promote gender diversity," the report stated.[13]

Overcoming barriers built up over generations of the energy sector being a male-dominated field won't be easy. Businesses will have to reconfigure benefits and programs that are tailored to men's needs but overlook women's challenges, including competitive maternity leave and family-friendly hours. Energy companies that do as much as possible to address these concerns will be the most effective at recruiting women. Those that allow women to contribute to business operations without having to compromise on their family responsibilities will take major strides toward eliminating the industry's gender gap. Women might look at energy companies differently when they see they offer fair maternity leave policies and safe spaces for breastfeeding.

So, as companies take steps to provide an environment that's fair and beneficial to all employees, male and female, they'll need to get the word out about those measures. This can be done as they advertise job openings, host interns, participate in job fairs, and take similar recruiting measures.

On our own continent, the Menengai Geothermal Development Project in Kenya has already completed a comprehensive gender assessment and is working to increase women's participation. Why can't more companies do the same?

Ending the Cycle of Harassment

African companies need to get serious about addressing and preventing sexual harassment in the workplace. Government leaders need to be part of the solution, too. I'm dismayed that as recently as 2021, more than a dozen African countries had yet to pass legislation that explicitly prohibits sexual harassment in the workforce.[14] We are failing 50% of our population.

We already know that women who work in male-dominated fields are considered more at risk of experiencing workplace sexual harassment.[15] That certainly includes extractive industries like oil and gas production and mining. The World Bank Group's report "Women's Employment in the Extractive Industry" described a despicable pattern of sexual violence and harassment where women, fearful of reprisals and losing their jobs, say nothing.[16] And if neither their company nor their government has policies and laws in place to protect them, I understand why women are afraid.

This horrendous cycle will continue until we stop it. We must enforce our laws to protect women. We must take a zero-tolerance stance on harassment in the workforce. And we must fill in gaps where legal protections are lacking. There is no valid excuse for tolerating violence and harassment of women on the job. Not for governments. Not for companies.

And if you have a leadership role at an energy company, addressing sexual violence and harassment in the workplace should be much, much more than an item on your to-do list.

This is much more than an environmental, social, or corporate governance (ESG) issue to be resolved. This is a moral issue.

I'm not asking for the impossible here. Look at the example of Malawi's power utility, ESCOM, which developed a "Social and Gender Inclusion and Anti-Sexual Harassment Policy" in partnership with U.S. agency Millennium Challenge Corporation (MCC). The 2018 policy calls for a culture of zero tolerance for sexual harassment and discrimination. It includes trainings to prevent gender-based violence in the workplace for board members, management, and staff. Companies can build on the example of South African mining companies, some of which have introduced hotlines for employees needing to report sexual harassment.[17]

We need to see these kinds of efforts become the norm. Companies need to provide staff sensitization programs about gender discrimination and prevention of gender-based violence; adopt strict guidelines on prevention; provide hotlines for employees to anonymously report sexual harassment and gender-based violence; and sensitize impacted communities about preventing gender-based violence.

Smoothing the Road for Female Entrepreneurs

The obstacles facing women aren't limited to the workplace. Women-owned companies have difficulty acquiring funding for their enterprises, inside and outside of the energy sector. The African Development Bank recently reported that the financing gap for women entrepreneurs in sub-Saharan Africa—including bank loans, angel investment, venture capital, and private equity—is more than USD20 billion.

The Director of Macroeconomic Policy, Forecasting, and Research at the African Development Bank, Hanan Morsy, recently wrote about the financing gap for the International Monetary Fund (IMF). "Credit rationing through high-interest rates disproportionately discourages women entrepreneurs from applying for loans, while lack of collateral can mean they have less access to loans than their male counterparts. And when they do have access, women typically face more stringent loan arrangements than men."

Women have a hard time accessing funding for their ideas, the founder of South African content and social media marketing start-up TRUE Content, Janine Jellars, told Africa.com. "Like most marginalized groups, we get funded based on our track records, not on potential. Often, in order to get the attention of funders, the burden is on us to prove our concepts beyond a shadow of a doubt. We also struggle to break that 'old boys network' and convince funders that 'women's ideas' are worth the investment."[18]

And in increasing numbers, women are reluctant to try.

"Fresh evidence drawn from the credit markets of 47 African countries suggests that women entrepreneurs in Africa, in general, and in North Africa, in particular, are more likely to self-select out of the credit market because of low *perceived* creditworthiness," Morsy wrote. "These women did not apply for loans or lines of credit because they were discouraged by their own perception that their applications would be denied."

Financial services providers should be a bigger part of the solution. One way to address the financing gap that disproportionately impacts female entrepreneurs is to create loans that accept smaller assets and non-traditional resources, like livestock, as collateral. Some banks in the West Bank and Gaza already offer innovative loan products for women, including collateral-free loans.

African leaders can also make financial literacy for women and girls a priority. Women entrepreneurs must be able to make informed financial decisions and take the steps necessary to increase their chances of being approved for financing. We also need more programs that empower female entrepreneurs, like the African Development Bank's Affirmative Finance Action for Women in Africa (AFAWA), which increases women's likelihood of obtaining financing through a risk-sharing mechanism. The AFAWA also provides training programs for female entrepreneurs, and it advocates for reforms intended to help women launch and grow their own businesses.

We also should be supporting the efforts of the African Women in Energy and Power (AWEaP) initiative, which is designed to accelerate African women entrepreneurs' participation in the power and energy sector. The program's core idea is that multi-stakeholder-driven initiatives that ensure the meaningful economic participation of women are key to wiping out African energy poverty.[19]

It's Time to Set Ambitious Goals

The African Energy Chamber is committed to leading by example in terms of collaboration and developing creative solutions to gender inequality in our industry. That is why we are a signatory to the global Equal by 30 Campaign,[20] a Natural Resources Canada-led initiative working toward equal pay, leadership, and other opportunities for women in the energy sector by 2030. Currently, we are among 144 organizations, partners, and governments who've signed on to support this campaign. Other African participants include my legal conglomerate, Centurion Law Group, along with East African energy consultancy Tsavo Oilfield Services, Equatorial Guinea-headquartered oil and gas service company Apex Industries, Mauritius-headquartered pan-African energy group DMWA Resources, the Germany Africa Business Forum, and Energy, Capital & Power. Each of us is committed to endorsing principles and taking actionable steps to address the gender gap in the energy sector.

As I said in May 2020, when the chamber became a campaign signatory, there is more urgency than ever to align political, financial, economic, and social resources to create a more equitable energy sector in Africa for all. We need to empower both to pursue academic backgrounds and training programs that will adequately prepare them for jobs in the clean energy transition.

As a coalition, we firmly believe that this cannot be done if women do not take an active role. Without women, there will continue to be a huge talent gap—Africa will struggle to fill available jobs in the energy sector. Preparing women ensures that companies will have the workers and leaders they need for the future.

As I have previously said,[21] the idea that there's no place for women in the energy industry is a myth and plainly wrong. There are many women already in the industry who have excelled. Women have started successful energy companies, serve as executive leaders in oil and gas companies, and spearhead innovation.

It would be tragic to usher in a bright, shiny era of renewable energy in Africa that leaves women on the sidelines. Doing right by women is not a side issue or icing on the cake. Women have to be an integral part of our energy transition because, quite simply, we cannot let half of our population miss out on energy industry opportunities. That would be immoral. And we can't talk about a just, equitable energy transition if we're willing to do that.

Isn't it time to give Africa its best chance at controlling the narrative and utilizing ALL of its available talent to become an international clean energy leader?

Chapter 14

THE VALUE OF COOPERATION AMONG OIL AND GAS STAKEHOLDERS AND ENVIRONMENTALISTS

Zambia's Luangwa River Valley is one of the richest wildlife habitats in Africa. It's home to one of the continent's most stable populations of elephants and also hosts African painted dogs, lions, leopards, and the rare Thornicroft's giraffe. Many of these animals find refuge in the network of national parks that have been established along the river and its tributaries, covering much of the eastern half of the country. All of these attributes make the valley an unparalleled treasure.

Despite these bastions of preservation, the area has also experienced high rates of deforestation over the last few decades.[1] It's not hard to understand why: agricultural and forestry resource management policies are insufficient, and there have been few efforts to expand the economy or the range of livelihoods open to local communities. As a result, the impoverished residents of these areas have been clearing forests for agricultural, charcoal, and firewood production.

To reverse this trend, Dr. Hassan Sachedina founded BioCarbon Partners (BCP) in 2012 to initiate and oversee reforestation projects under the United Nations' Reducing Emissions from Deforestation and Degradation (REDD+) program. One of BCP's early projects was the Luangwa Community Forests Project (LCFP), an effort launched in 2014 that would eventually cover more than one million hectares (nearly 2.5 million acres) of land and involve more than 170,000 people in the local communities. The LCFP has become Africa's largest REDD+ project and earned gold level validation in all three categories for Climate, Community, and Biodiversity Standard.[2]

In short, the project is one of Africa's brightest environmental stars.

One of the contributors to BCP's success has been Eni, the Italian major. Eni is active in 66 countries around the world, and its upstream portfolio includes exploration and development assets in 14 different African states.[3] But Eni isn't just about oil and gas. In recent years, it has also devoted an increasing amount of time to the energy transition, promoting cleaner forms of energy and participating in decarbonization projects to offset the environmental effects of its more traditional energy pursuits. It has set for itself the ambitious goal of achieving zero net upstream emissions by 2030, and it aims to compensate for its residual emissions through the support of forest conservation.

As part of this zero net upstream emissions initiative, in 2019, Eni signed on to become an active member of LCFP's governing board in Zambia. It committed to purchase carbon credits certified under the Verified Carbon Standard and Climate, Community, and Biodiversity Standard for the next 20 years. Proceeds from these purchases will go directly to support forest management activities and community improvement projects in the 12 chiefdoms encompassed by the LCFP.

The benefits of this partnership became evident almost immediately. Less than a month after Eni joined the board, the Luembe district administration embarked upon a project that will improve access to education for more than 380 families, renovating two classrooms and completing a home for teachers living in a remote region along the Lukusashi River. Elsewhere in the district, BCP has also been able to invest in a hammer mill and provide boats for improved river access in areas with no roads.

The impact of these measures is impossible to ignore. In some of the communities where BCP has worked for at least five years, household incomes have increased up to 400%. Likewise, wildlife surveys covering some areas of the forest also show up to a 300% increase in the conservation of high-value species within the last five years, as well as the re-establishment of important carnivore species.[4]

This is all remarkably good news. But it's not the whole story. It also shows what environmentalists and oil companies can accomplish when they work together toward a common goal.

False Perceptions

All too often, those who are passionate about the environment see oil and gas companies as ruthless, pursuing profit at all costs and showing no interest in protecting the environment. On the flip side, those involved in the oil industry often fall into the same pattern of negative and overly generalized thinking about environmentalists, accusing them of having tunnel vision or being more interested in the planet than people.

This zero-sum thinking is perhaps understandable, but it's also unproductive—because there are other options. Eni and its collaboration with the LCFP provide an example of what is possible when oil and gas stakeholders and environmental organizations meet each other in the middle.

That's why I applaud the United Nations' REDD+ program, which was specifically designed to help environmentalists engage with companies like Eni in meaningful and productive ways. Choosing the REDD+ format for their project gave Dr. Sachedina and BCP a way to make their project more welcoming to corporate participants such as Eni. But it also allowed them to do so without tipping the balance of power, as often happens with corporate patronage. Instead, representatives of two sides that are often diametrically opposed were able to establish a safe transactional structure that allowed both to benefit.

I hope we'll see many, many more examples of this kind of cooperation. Instead of working at cross purposes, oil and gas industry stakeholders—including African governments and petroleum companies—and those devoted to protecting the environment should work together to find common ground, shared goals, and ways to support one another's efforts for the good of Africa.

These collaborations will become increasingly important as African states strive to achieve the emissions-reduction objectives outlined in the Paris Agreement on Climate Change and grow their energy mix. These goals will not be easy or cheap to reach, especially for developing countries—if oil and gas companies are willing to do the work and foot the bill, we ought to make it easier for them to do so.

Think about it: How else are countries that rely on profits from fossil fuels for much of their revenue supposed to navigate this transition? What about countries struggling with energy poverty? What about countries that are just beginning to put policies and programs in place to capitalize on their oil and gas resources? Can they really overcome their challenges and meet their commitments to the international community if we don't take an all-hands-on-deck approach?

Instead, let's look for ways to extend the REDD+ program and scale it up.

Finding a Way Forward

Of course, one shining example does not a paradigm shift make. But we seem to be on our way. More and more, international oil companies (IOCs) are supporting environmental initiatives, which increasingly are being recognized as smart business practices—as means of ensuring long-term profitability and building goodwill.

For example, Tullow Oil of Ireland, which promotes itself as an advocate of sustainable development, began touting its support of the Paris Climate Agreement on its website in 2018. Unlike Eni, Tullow doesn't sponsor conservation projects to offset its emissions. However, the company set a series of targets for actively reducing its emissions and was able to make real progress after just a year of action. Although it increased its overall CO_2 emissions by 3.2% from 2018 to 2019 (which the company attributes to drilling campaigns with the Stena Forth and Maersk Venturer rig, seismic and exploration activity in Guyana and Comoros, and the Early Oil Pilot Scheme in Kenya), Tullow lowered its emissions intensity from 139 tons of CO_2 per 1000 tons of hydrocarbons produced to 134 tons per 1,000, a decrease of 3.6%. Much of that reduction is due to reinjecting natural gas that normally would have been flared into a reservoir at one of the company's major operations in Ghana.[5]

Tullow also established a Biodiversity Advisory Panel in 2019 to review its activities and manage impacts from its Kenyan Joint Venture operation, which overlaps two World Heritage Sites around Lake Turkana and the Great Rift Valley. The panel is responsible for making sure management and mitigation plans are flexible enough to respond to changing data and requirements, coordinating with local and national government on-site rehabilitation, and integrating project activities with local development programs to improve the livelihoods of residents.

Getting Ahead of the Problem

Corporate innovation of this kind is good to see, but it's not enough on its own. Nor is corporate collaboration with the non-profit arena.

To balance resource development and conservation, governments will have to play a part, too. They can't leave things to chance, since lax oversight, weak regulations, and poor disaster preparation in developing countries can lead to environmental catastrophes that give development efforts a black eye and take decades to clean up.

Ecuador can testify to that. In the 1970s, the country's state oil company built a pipeline to connect its oilfields in the Amazon River basin with ports on the Pacific coast. The pipeline had to cross through the Andes Mountains, a range that includes a number of active volcanoes and is bedeviled by frequent earthquakes and heavy rain on its eastern slopes. Through a phenomenal lack of foresight, the state oil company chose to route the link through a river valley right beneath the foot of a jungle volcano known for centuries as El Reventador ("The Exploder"). Ever since then, the pipe has suffered dozens of ruptures and spills because of earthquakes, volcanic eruptions, and storms along with the massive landslides and extensive flooding that often follow them. One of these incidents destroyed dozens of miles of pipe in 1987.[6]

While this scenario could be considered an example of the resource curse, disasters caused by oil and gas operations are not the norm—and they don't have to be. One of the key things that will allow African countries, and their people, to benefit fully from their oil and gas resources is a commitment to good governance. That includes carefully negotiating exploration and production contracts with oil and gas companies and including measures to protect local environments.

One of the ways government leaders can achieve these benefits is to consult with experts from environmental organizations and governments with effective oil and gas policies—and there are plenty of them out there. That's what Mozambique has been doing.

Mozambique is projected to become the world's third-largest natural gas exporter as soon as 2023, with gas projects contributing an estimated USD39 billion and up to 700,000 jobs to its economy over the next two decades.[7] To better prepare itself for this welcome development, the government of Mozambique commissioned a report in collaboration with the U.N. Environment Program (UNEP) and Norway's Oil for Development Programme to create a roadmap for managing environmental challenges associated with the oil and gas sector.

The report was released in Maputo on March 15, 2019, at an event attended by representatives of multiple government ministries, international development organizations, and academic institutions, as well as delegates from ExxonMobil, Anadarko, and Sasol, some of the largest IOCs working in Mozambique. The document recommended that the Mozambican government take action on the following fronts:

- Updating oil spill preparedness and response strategy on both land and sea.
- Establishing a mechanism to monitor, document, and regulate emissions, particularly methane, from major industries.
- Increasing the resources and technical capacity of ministries charged with oversight.
- Developing more specific regulations and operational procedures to improve coordination and communication between ministries and reduce overlapping roles and responsibilities.

By the time this report was published, Mozambique was already relatively farther along in its resource development than many other African countries. But it was not so far along as to make course corrections unnecessary. The government's willingness to recognize the pitfalls ahead and to begin actively planning to combat them is an admirable example of responsible governance.

Gathering Local Input

While Mozambique gained valuable insights and knowledge from working with the UNEP and Norwegian petroleum industry experts, relying on "outside" resources shouldn't be the only option. Homegrown expertise, communication, and cooperation are just as important. We only have to look as far as Nigeria to see what happens when input isn't solicited or accepted at the local level.

Nigeria has a history of adopting industry management strategies first developed in other countries such as the United States at times when it lacked the technical capabilities or expertise to develop its own. What's more, Nigerian authorities have also tried to apply the same strategies in every part of the country. But this copy-and-paste approach never really worked, as it took no account of local sociocultural, environmental, or economic norms. As a result, efforts to develop an effective policy regime have often been too little, too late, and too deeply compromised by corruption to win support from the communities in the Niger River Delta most affected by the environmental consequences of oil and gas development.

As a result, members of these communities have lost faith in the government's ability to handle their problems. They don't believe that anyone in Abuja cares about the oil spills, pipeline leaks, and chemical releases that are affecting the places they live.

That's an understandable response, given the failure of past policy initiatives. But things are turning around. Or at least that's the intention, as Nigeria is now employing a business methodology known as "stakeholder engagement" to help develop its environmental policy.[8]

Within this framework, the government actively seeks input from stakeholders—that is, the people who have a direct stake in the outcome of policy and business operations—so that it can start to rebuild trust, promote transparency, and underscore its legitimacy in the eyes of its citizens. Stakeholders are not a uniform group, as they range from residents of the areas near production facilities to company employees to government monitors and officials.

First and foremost, though, this approach requires Nigerian authorities to treat stakeholders as people, and not as abstract conceptual groups.

This approach makes sense on a philosophical level, as it respects stakeholders as individuals with their own unique goals and values. But it's also practical since Nigeria's government already knows that the one-track approach it's taken in the past succeeded only in destroying trust quickly.

Nigeria is simply not the kind of place where a single overarching strategy for community involvement can be expected to work. After all, the country encompasses hundreds of tribal groups that vary in their levels of local autonomy, community structures, and even the languages they speak. As a result, a lot of stakeholders co-exist, and many of them have competing aims. So the challenge has been to not only find out what sociocultural, economic, and environmental priorities these communities value, but how decisions to prioritize them should be made—and how to incorporate these considerations into effective environmental policies that could make the Niger Delta a better place to live.

To this end, Nigeria has been surveying stakeholders—and tailoring its surveys to various communities based on language, communication style, education level, and local understanding. In locations where efforts to solicit input by email or telephone were impractical or impossible, government data collectors hit the road to perform face-to-face interviews in villages and towns to gather the maximum amount of data from the public. In addition to community members, the government also surveyed three disparate groups of stakeholders: experts, regulators, and operators.

The results of this information-gathering campaign revealed that, in many ways, different groups of stakeholders thought alike. Each was acutely aware of the extent and impact of oil and gas pollution in the Niger River Delta region, as well as the effect it had on both human and economic health. They also said their top three priorities were drinking water quality, agricultural soil quality, and food supply safety. Prescriptions to remedy these problems have included giving the government stronger enforcement powers, implementing existing policies, and providing a system for faster and easier reporting of spills, accompanied by a prompt and thorough response and greater involvement of citizens in the reporting process.

Since the government was acting in its capacity as the legal owner of the country's hydrocarbon resources, this campaign was, in effect, an instance of corporate social responsibility (CSR) in action. And that's a good thing. CSR and stakeholder engagement aren't just buzzwords; they're tools that can entice hydrocarbon companies to work *with* the government instead of *around* it. If so, they should give the Nigerian government an incentive to maintain discipline on this front so that it can turn these early efforts into long-lasting results.

In other words, a good foundation has finally been laid for effective management of Nigeria's oil and gas resources.

Profit Versus Purity

So where do we go from here? How can we ensure that moves toward cooperation between oil companies, environmentalists, governments, and stakeholders continue, and that new patterns of collaboration emerge?

Compromise is key. With compromise, oil companies can put people ahead of profit, and environmentalists can put people ahead of the planet. If progress is to continue, each side needs to stop demanding absolute purity from the other.

Is it appropriate or pragmatic to say "keep it in the ground" to an oil company when there's no substitute for fossil fuels as a life force for the developing world?

Such arguments most often come from well-to-do countries where the lights always come on when you flip the switch, and it's easy to get wherever you want to go. It's easy to assume there are alternatives when you have had reliable electricity and transportation for generations.

But where does that reliability come from? Not from new technologies on the cutting edge, but from old, tried-and-true technologies with wide distribution and support—such as the gas-fired power stations that are still necessary to compensate for the natural fluctuations in renewable energy generation. No matter what exciting possibilities the future holds, fossil fuels are still the backbone of the current world energy economy—and a major requirement for developing economies, where access to fuel and electricity cannot be taken for granted.

Certainly, there are already some promising wind and solar generation projects on the African continent, and we should continue to build on that! However, these schemes are still not capable of providing enough power to meet Africa's needs.

Additionally, Africa has no means of producing solar cells and wind turbines for its own use.

Currently, almost all of the technology and expertise needed to make these technologies must be imported to bring these projects to fruition. This may change in the future, especially if the continent can secure investment for the development of its minerals, metals, and rare earth resources, all of which are needed to support renewable energy generation.[9] However, the payoff for such efforts is far down the road—and it's not even clear exactly how far.

By contrast, oil and gas lie under African soil *now*. The technologies and equipment needed to extract them from the ground have already been developed, and production facilities exist on African soil *now*. Africans trained in the necessary science and engineering disciplines are available *now*. African workers with experience in the industry are ready to work *now*.

Other countries have spent the last 200 years exercising their rights to exploit resources within their own borders (not to mention that of their colonies) to build a modern world for themselves. Should Africa be denied those same opportunities now because the old technology is now out of fashion?

The old saying goes that perfect is the enemy of good. Clean energy is a wonderful ideal, despite its imperfections, and it may very well be achievable—but right now, it's not for everybody, and it's not possible today.

Environmentalists can continue to fight for clean air, soil, and water and still acknowledge that people who live in the dark do not want to continue waiting in the dark—and shouldn't have to do so—until the perfect power solution comes along. If a village in rural South Sudan has a choice between a solar power plant in 20 years or a gas-fired plant in 2, which do you expect them to choose? It makes sense for them to work with what they have today, for the future is promised to no one.

The good news is that Africa's future doesn't have to be everyone else's past. There is another way. Let modern clean energy solutions spread organically across the landscape on their own merit—and let environmentalists continue to work with fossil fuel producers to improve sustainability, promote the best possible energy mix, and create a balance that benefits people through progress and profits.

Chapter 15

USING WISDOM FROM OTHER PARTS OF THE GLOBE TO TRANSITION TOWARD AFRICA'S NEW ENERGY MIX

According to Nigerian folklore, three friends—Food, Wealth, and Wisdom—began a journey one day to search for a place to live. On the first day of the journey, they came across a man who was sitting against a tree. They told the man that they were looking for a home, and the man said he would be happy for Wealth to live with him. But Wealth rebuked the man, saying that if he had invited Wisdom to live with him, all three of them could have moved in. If Wealth moved into the man's house by himself, he wouldn't be around for long without Wisdom.

Soon, the group came across another man and told him of their search to find a place to live. The man promptly invited Food to live with him. Food told the man that if only he had chosen another one of them, he could have had all three of them live with him. But because he chose the wrong one, he would not be able to keep Food around for long.

Finally, the group came across a third man who was at work. When he learned of the group's journey, he said that he would like Wisdom to live with him. Food and Wealth quickly spoke up, telling the man that they too would move in with him because they knew that Wisdom would take good care of him.[1]

There's a reason that people still tell and re-tell old folktales. These stories, handed down through generations, typically portray proven knowledge and logic that can be applied to modern circumstances.

The story of the three friends and their search for a home reminds me of Africa and its journey toward an energy mix in line with current global demands: Just as the men in the story learned, we must learn to put Wisdom first—especially when considering Africa's transition to an expanded energy mix. That's why it's worthwhile to look to other countries to learn what they did right—and wrong—to make the journey easier.

We must find a path that will best suit Africa's unique circumstances—in part, electricity for hundreds of millions of people lacking electricity could be powered by our vast natural gas resources, while other African countries currently rely heavily on oil revenue for their budgets.

We can learn from other countries; however, it's important to remember that we can't simply apply their strategies as one-size-fits-all solutions.

In a paper published by Taylor and Francis, the authors argue that policy responses to climate change should ensure that justice and fairness are upheld, and that because of the differences across countries, we should "urge the development of clusters of models appropriate to jurisdictions with similar varieties of political and economic institutions and economies." The authors conclude by saying that jurisdictions' ability to "steer complex, long-term transitions are therefore a crucial variable in transitions success."[2]

And that brings me to my point: much like the men in the Nigerian folktale, Africa must invite Wisdom to live with her as it looks closely at and learns from what other countries have done to achieve their energy goals. At the same time, though, we must also keep in mind its differences and evaluate the practices in light of its own unique needs.

Now, let's take a look at what some other countries have done when transitioning to renewables and what Africa can learn from them.

Alberta Builds on Its Strengths

Like Africa, the province of Alberta, Canada is rich in natural gas. Gas has made up a huge portion of Alberta's energy economy for more than 100 years. In the 2005-2006 fiscal year, Alberta produced 14.1 billion cubic feet per day of it and collected CAD8.4 billion in gas and by-product royalties.

Fast forward to 2019, and things looked very different: Alberta was forced to sell gas at a discount due to such factors as increased competition from the U.S., minimal pipeline access, and construction delays when the country attempted to expand its pipeline infrastructure. The result: a dramatic decrease to CAD371 million in royalties from selling 11.1 billion cubic feet per day.[3]

So why is it worth looking at Alberta as an example for Africa's transition? Because of what it did next. Rather than abandoning its rich pool of resources, Alberta pivoted and reimagined how it could use those resources to build jobs, strengthen its economy, and provide valuable commodities to the world.

Alberta began the process by looking at its strengths and imagining how it could use them to ensure the long-term success of its natural gas sector. It determined that its strengths were:

- An advanced and innovative upstream natural gas sector that used service and supply sectors to drive technologies that could reduce emissions by capturing and storing carbon.
- Low production costs.

- Abundant natural gas liquid resources that could be converted to petrochemicals, used in oil sands production, and harnessed for power generation and hydrogen production.
- A strong track record of protecting the environment.
- Superior petrochemical production.
- Natural gas infrastructure that was sufficient to provide for value chains.
- A skilled workforce.
- Innovation funding that provided for energy technology and research.
- Established connections that provided access to markets.

Understanding that natural gas will play an important role in the world's energy needs for some time to come, in 2020, Alberta released its *Natural Gas Vision and Strategy*. It calls for utilizing natural gas to further grow and diversify Alberta's economy, whether that involves encouraging the development of more petrochemical plants that use natural gas as a feedstock or using natural gas to grow Alberta's plastics recycling sector. Not only will the province's strategies result in economic growth and job creation in the short term, but they also will lay a foundation for long-term growth that will benefit Canada as well as the global market.

Here are the stated goals of the province:

- To resolve the system barriers to attract investors and move gas to markets.
- To drive investments in the natural gas value chain to enable growth.
- To deploy natural gas into new markets, including LNG, petrochemicals, and hydrogen.
- To improve value-chain integration using research and development along with new technologies.

- To build alliances with industry, other Canadian provinces, and the federal government to ensure that Alberta's natural gas and value-chain products can compete globally and attract international investors.

Alberta could have bowed to international pressure to walk away from fossil fuels and left its natural gas resources in the ground. Instead, the province showed true wisdom and built on what it has by turning a portion of its natural gas into something the world needs—and will continue to need in the future.

And Alberta won't need to rely on aid to make the transition. Part of the wisdom of this plan is that Alberta will attract investors who might not have otherwise been inclined to invest. Because Alberta is changing the way it uses natural gas, the stigma that is often attached to oil and gas production is removed.

So how can we apply this wisdom to Africa?

First of all, it's important to notice that Alberta doesn't use its natural gas reserves to simply supply other countries. Instead, it will utilize its resources to manufacture new goods (LNG, petrochemical manufacturing, plastics recycling, and hydrogen) and supply much-needed goods to the world.

To date, Africa has supplied oil and gas to the world—while neglecting its own needs. Instead, African countries should understand the power of the resources they have and make a plan to use them to build their economies. Then, they'll have the revenue and economic diversity necessary to ensure that their energy transition improves citizens' lives, rather than subjecting Africans to the job losses, economic hardship, and unrest that a rash, premature exit from oil and gas activities would likely cause.

Where should the continent begin? If African countries are to learn from Alberta, they should begin by taking stock of their strengths. Here are a few to consider:

- An abundance of oil and natural gas—and more being discovered all the time.

- A large pool of young workers who are willing and able to pivot and adjust.
- Leaders who understand the importance of environmental responsibility.
- Some established connections.

Next, African leaders should evaluate these strengths and build on them to make a plan. Here are a few suggestions:

- Use offshore oil and gas to provide returns to African states to help them transition.
- Identify, build, and scale up profitable projects that offer low-carbon energy solutions.
- Continue to develop technology such as carbon capture and storage.
- Be persistent in asking oil companies, "What is your strategy for low-carbon energy production and the energy transition, and how can we work with you on it?"
- Explore and develop the petrochemical industry. For instance, in Africa, producing fertilizers goes hand-in-hand with our ongoing need for successful farming.

You get the point. Africa needs to use its strengths methodically and knowledgeably, and not in a way that helps the rest of the world and ignores its own needs. No, this time, Africa must look long-term and plan its actions in a way that will allow it to stand on its own two feet sometime in the near future.

By adapting Alberta's approach to Africa—and by striving to become a manufacturing powerhouse instead of simply a supplier—we can achieve that goal. In other words, African countries need to think about building local economies that will provide jobs, security, and a future for their people.

India: Taking One Bite at a Time

India is growing quickly, and as a result, its consumption of oil and gas is also increasing. From 2007 to 2017, India's total fuel consumption increased by 50%. But the country has lofty environmental goals: by 2030, it plans to increase its use of non-fossil fuels for electricity generation to 40%.

What makes this strategy especially notable? India is doing it one small step at a time.

For instance, India has reached 100% electrification for its citizens, and it did so while introducing centralized and distributed renewable energy sources in the process. In other words, rather than transitioning in one fell swoop, the country introduced these renewable sources little by little as it built and improved its grid. While the country is still reliant on fossil fuels for more than 60% of its energy,[4] these small steps are driving it toward a future where renewables will play a major role. By instituting technology improvements and economies of scale, it managed to reduce the cost of solar tariffs from US10 cents a kWh in 2015 to US3.57 cents a kWh in 2020.

Another small but important step is India's transition to LED bulbs in residential homes and streetlamps.[5] In just five years, India expanded its use of LED bulbs 130 times over. That seemingly small step saves more than 40 TWh of electricity a year. To put these figures in perspective: The amount of electricity saved could power 37 million average Indian households a year, or the entire country of Denmark! What's more, India, which had only a fraction of the global LED market before, now enjoys 10% of it.

Finally, India is demonstrating true wisdom by collaborating with other counties and private sectors for the development of sustainable energy.[6] For instance, to attract foreign collaborators, India and the U.S. have joined forces to create the Partnership to Advance Clean Energy (PACE). Under this agreement, when the two countries join together to facilitate research and development projects in renewables, they share the financing of the project.

Through this unique collaboration, the two countries are exploring the feasibility and eco-friendliness of India using natural gas technology to create a more stable and reliable energy grid. Because the U.S. is flush with natural gas reserves, this collaboration could be a boon to both countries.

The significance of this move? Using the philosophy of one of Africa's oldest folklore tales, India invited Wisdom to live with it, and the results are promising.

Will Africa do the same by seeking out the private sector and government-to-government collaborations to facilitate its transition? I certainly hope we consider it.

But this is not the only thing Africa can learn from India's ambitious energy plans. We should also take note of India's strategy of making changes in small steps rather than one large leap. Africa hasn't even achieved 100% electrification yet, so the thought of taking a giant leap to 100% of renewables simply isn't realistic.

But it *is* realistic to adapt India's strategy of slowly integrating renewables into area-specific grids.

And finally, much the way that India and the U.S. forged a strong, mutually beneficial partnership, African countries can also collaborate with other nations. Lest it forgets its own advice, inviting Wisdom into your home can only result in good things.

Taiwan Runs Like the Wind Toward Renewables

Many countries have made well-intentioned plans when it comes to adding renewables to their energy mix, but not all have met them. Enter Taiwan, a leading offshore wind market in East Asia because it has been able to accomplish so much in a short amount of time.

For instance, Taiwan increased its wind energy target to 5.5GW by 2025 in January 2018. And it was so successful that it increased its target again to add 10GW offshore wind capacity between 2026 and 2035.[7]

This success has come without serious efforts on the part of Taiwan's government, which understood early on that it couldn't rely on the low costs enjoyed in already mature energy markets. Instead, it worked to generate its own volume in developing infrastructure, the local workforce, supply chains, and project finance. By planning and building the components necessary to support long-term growth, the government ensured that it had a basis for cost reduction in the long term.

Remember our African folktale? If Taiwan had instead tried to rely on mature markets for analysis of long-term cost reduction (immediate Wealth), it would have been going against the wind. But it didn't. Instead, it used planning and forethought (Wisdom) to ensure that its wind projects would be successful now and into the future.

Another way that Taiwan's planning paid off was in the zoning of offshore wind farms. Rather than exposing developers to the risk and increased costs associated with zoning clashes, Taiwan carefully studied and mapped out zoned regions for developers to bid on. This not only sped up the development process, but also made the environment much friendlier to developers—and investors.

The race to wind power in Taiwan was almost brought to a halt in 2018 when the government decided to dramatically slash the Feed-in-Tariff level. The Global Wind Energy Council (GWEC) met with the Taiwanese government to explain how such a reduction in tariffs would stall the market, and for months they worked tirelessly to come up with a solution that would benefit everyone. What could have been a major hiccup in Taiwan's wind energy plan led to a better financial design, but only because the government was willing to listen to and collaborate with outside sources.

Now, let's travel across the Indian Ocean to Africa.

Africa, like Taiwan, has great potential for offshore wind energy, especially in east Africa, where strong winds blow in from the ocean. In fact, according to the International Renewable Energy Agency (IRENA) *REmap 2030* report, Africa has the best wind energy resources in the world with a potential of 100GW. And while some progress has been made, like the wind power plant that launched in Senegal last year,[8] there is much more to be done.

Anyone who has seen the Netflix original movie *The Boy Who Harnessed the Wind* understands just how impactful wind power can be. The true story takes place in Northern Malawi, a poor African village that lacked water for its crops due to drought. A young boy came up with the brilliant idea to harness wind energy using spare bicycle parts to bring water to the crops and saved the day. Pure genius!

But including wind power as a major source of energy in Africa will take a lot more. In fact, we can look directly to Taiwan for what Africa needs to continue its transition to renewables.

And we can look at South Africa as an example: Many say that the country's Renewable Energy Independent Power Producer Procurement Programme (REIPPPP) is behind its success when it comes to wind power. The country also works with developers who put effort into using new tools and resources. Plus, the government is paving the way to make investing and partnering with South Africa even more convenient. For instance, it launched the Wind Atlas for South Africa to aid developers in understanding African wind patterns.[9]

If countries like Ethiopia, Ghana, and Kenya—which, according to the World Bank, have the most potential to develop wind power—would open the lines of communication with South Africa and learn from its successes, they would have a better chance of improving the energy grid in their countries.

African countries understand that they need to expand their energy mix with renewables, and they have the resources to do it. They simply need the time to map out strategies and make sure Wisdom guides their efforts.

It's time for Africa to rise up and take its place in the world. But that will take an arduous amount of planning, foresight, and collaboration.

I, for one, believe that Africa has what it takes to finally realize its potential. If we act with Wisdom, other benefits—such as increased Wealth and resources—will surely follow!

Chapter 16

SUCCESSFUL AFRICAN ENERGY ENTREPRENEURS

When Vere Shaba was a secondary student, she saw a satellite image of Earth at night. While most people are drawn to the glow of city lights captured in such images, Shaba immediately noticed something else.

"My first question was, why is Africa so dark? How are our children reading at night?" Shaba asked.[1]

And if children couldn't read at night, she worried, how could they possibly complete their homework?

"If education is the greatest weapon that we can use to change the world, how will we ever be empowered as a continent if the education of African children stops when the sun sets?" Shaba asked herself.[2]

This is a valid question: In 2022, approximately 600 million people in Sub-Saharan Africa live without electricity. And even if you happen to live in a part of Africa that does have power, reliable service isn't a given. Grid blackouts are common, and you're likely to lose power just when you need it the most.

Put another way: The supply of energy in Africa has not kept up with the increasing demand for it.

So, what is it like for children to try to read at night without access to electricity or light? Some don't even try, while others read by dangerous open flames or kerosene with carcinogenic emissions. Something so basic and taken-for-granted in many parts of the world has deep ramifications for people in Africa who go without!

But some aspiring African entrepreneurs are working to solve this problem—and to ensure that African children have access to safe, reliable power sources. You may not be surprised to learn that Shaba is one of them: She decided to study engineering so she could "engineer this problem out of the continent."[3] This became Shaba's calling, and she set out to try and tackle energy poverty and issues of sustainability. She studied mechanical engineering at the University of Cape Town, and when she graduated, she became the founder and CEO of a Johannesburg, South Africa-based consulting engineering firm called Greendesign. She was recently featured on the cover of *Forbes* magazine.

I think Shaba's story is an excellent example of how curiosity, commitment, and determination can help young Africans fulfill their callings.

African entrepreneurs like Shaba are finding opportunities in the renewable energy space. Studying their successes, and even their failures, will help more African energy companies launch and thrive. We can grow Africa's energy mix effectively and make sure renewable energy provides African economies with many of the same benefits that oil and gas can deliver, including job growth and knowledge-sharing opportunities.

Barriers for African Energy Entrepreneurs

Becoming a successful entrepreneur is never easy, but starting a business can be particularly challenging for Africans. What might seem like a small, manageable obstacle to a Westerner may be a substantial, insurmountable roadblock for someone starting a company in the African energy space.

Here are just a few of the hurdles Africans are forced to jump:

Limited Access to Financing

Lack of access to capital is the #1 barrier facing aspiring African entrepreneurs. This is especially true for small and medium businesses in the renewable energy sector. Investors consider off-grid energy projects like wind farms that sell energy directly to consumers risky. Banks are not eager to invest in new and unfamiliar ventures.

On top of that, many African would-be entrepreneurs do not even have a credit report, let alone the stack of financial statements and documentation their Western counterparts can produce to demonstrate their ability to operate successfully. When investors, development institutions, and hedge funds base financing eligibility on the same criteria for Western and African candidates, Africans are automatically at a disadvantage.

And because many African countries do not have access to capital, most of the financing comes from Western institutions, governments, and organizations. Yes, African entrepreneurs can attempt to form joint partnerships with Western companies, but that has its limitations, too. African companies negotiate partnership agreements from a very weak position. In most cases, their Western counterparts are coming with the money, and an African government is giving them a concession to build wind and solar projects in the African entrepreneur's position. So, instead of growing as an entrepreneur, the African company becomes more of a commission agent facilitating foreign capital, government concessions, and local labor. It's called a joint partnership, but the African company plays a subservient role.

Lack of Technical Knowledge

Many clean energy start-ups require a great degree of technical knowledge. What we're seeing, though, is an education and training deficit among young Africans, limiting their opportunities for success as employees and as entrepreneurs. Consider solar mini-grids: they represent opportunities for technicians who can install, maintain, and repair them. But the number of people qualified to do the work is not what it should be. According to a report by the Energy and Environment Partnership Trust Fund (EEP Africa), a clean energy finance facility, Kenya only has 2,100 certified engineers out of a population of 45 million.

"Some mini-grid developers are addressing this issue by collaborating with vocational training institutes to recruit technicians," EEP Africa wrote. "Others are setting up their own local or online training programmes to upskill rural mechanics."

These training efforts are encouraging, but we'll need many, many more in order for Africa's rapidly growing population to compete effectively for job and entrepreneurial opportunities.

Minimal Business Experience

The majority of people living in Sub-Saharan Africa are under the age of 25. We're seeing more than 10 million young people entering the job market each year while job growth remains stuck at around 3 million new positions annually.[4] The lack of available jobs has nudged many young Africans toward entrepreneurship.

In theory, this is a good thing: Young people are creating more jobs. However, this segment's lack of business experience can lead to missteps if young entrepreneurs lack access to advice and consulting from more experienced entrepreneurs. They simply don't have the business management experience that organizations often need.

Successful African entrepreneurs can offer support by serving as mentors, and African companies can make a difference by offering young people internships to cultivate the soft skills necessary for business success.

Governmental Differences in Africa

Transitioning to green energy requires an enormous amount of capital investment from governments to succeed. This includes investing in entrepreneurial projects. Unfortunately, governments throughout Africa have varying perspectives and priorities when it comes to the energy transition and support for African-born businesses (in reforms or regulations). These governments also have varying access to capital. For instance, in Morocco, Law 13-09 liberalizes renewables and makes it easy for entrepreneurs to export renewables and attract investors.[5] But we aren't seeing enough policies like these.

The State of Energy Entrepreneurism in Africa

It is difficult to quantify exactly how many African entrepreneurs exist within the energy space. I am seeing some start-ups, particularly in the cost-effective solar solution space. From 2015 to 2018, the latest data available before the devastation of the COVID-19 pandemic, solar start-ups were the most likely to attract investors.

Writing for *The Beam Magazine*, authors Akinyi Ochieng and Fadekemi Abiru highlighted a number of promising initiatives that, despite the challenges facing African entrepreneurs, succeeded in getting funding[6]:

- In 2017, pay-as-you-go energy provider M-KOPA secured USD80 million for solar installations.
- In 2016, Lumos secured USD90 million for pay-as-you-go solar solutions in Nigeria.
- In 2016, BBOXX Ltd. raised USD20 million.

- In 2016, d.light Energy Pvt Ltd raised USD15 million.

Ochieng and Abiru argued that solar is the fastest and cheapest solution to Africa's energy crisis. Although I agree that solar is very important, I think it is also important to remember that solar is only one facet of a diversified and multi-disciplined solution. It alone will not be enough. Too many people need power, and the infrastructure to provide it is nearly nonexistent.

We need more entrepreneurs throughout the energy space to step up, innovate, and develop new opportunities. And we need government regulations that foster business success for Africans and ensure that the African people get their fair share of the rewards. This requires change and investment—investment in our people and training, investment in African-grown businesses, and investment in the infrastructure required to become a major player in the global energy space.

The rise of mobile phone usage in Africa has only increased the demand for power throughout Africa.[7] Yet, once more Africans have regular access to the energy that they need, the limits on our young people's potential can be lifted. With internet and power, mobile phones, and laptop computers, Africans can more readily work remotely, maintain their incomes during pandemics, and better provide for their families.

I have the privilege of meeting successful African entrepreneurs every day throughout my work in Africa. I can see the impacts they're having on their communities. More investments and opportunities will continue in this space. I can't wait to empower and meet the future entrepreneurs that will bring Africa closer to 100% access to electricity and give all Africans a better chance at the future we all deserve.

Inspiration and Impacts

Henri Nyakarundi knows firsthand about starting a business for the purpose of helping others. His company, African Renewable Energy

Distributor (ARED), harnesses solar power to meet internet connectivity and phone-charging needs in Rwanda, Uganda, Nigeria, and Côte d'Ivoire. By bridging digital and energy gaps in Africa, Nyakarundi says, his company is playing a role in fighting poverty. "I wanted to do something that will not just focus on making money, but that was going to have an impact on people, especially on the low end of the (income) spectrum," Nyakarundi said during a 2021 phone interview.

Nyakarundi, who was born in Kenya and raised in Burundi, studied computer science at Georgia State University in the U.S. He got his start as an entrepreneur in 2006 and 2007, when he founded trucking and logistics companies UMG Transportation and UMG Logistics in Burundi and America. He used the revenue from those ventures to pursue his vision for ARED in 2013.[8]

"My previous businesses, I didn't necessarily like the business, I just liked the fact that they could generate money," Nyakarundi said. "I love ARED because it really addresses some of the key points I've always wanted. I've always wanted to start something, develop my own product, something unique. I wanted to solve a huge problem, which was connectivity and access to energy, specifically for charging. And I wanted to have an impact: I wanted to solve a problem that nobody's really focusing on."

ARED offers franchisees the opportunity to lease solar-powered kiosks. Each one can charge more than 30 devices at a time and provide internet access and a Bluetooth printer.[9]

While ARED is thriving today, Nyakarundi overcame significant obstacles to make the company a success. Developing a product-based business is different than launching a service-based one, he said. It requires more capital, product development, and support. He invested large amounts of his own money to develop prototypes for a solar-powered charging kiosk (it took three tries to get the results he needed), and he ran out of funds more than once during the complex product development and testing cycles that followed. Fortunately, he was able to win support for his business's vision and obtain the necessary financing to move forward.

So, what advice would Nyakarundi offer other Africans thinking about starting their own energy companies?

- **Be flexible and multifaceted.** To build a sustainable business model, you'll need to solve multiple problems with one platform, Nyakarundi said. That's why ARED kiosks meet connectivity and phone-charging needs.
- **Understand that you're going to go through a lot of money as you develop your technology, fine-tune your business model, learn to minimize your business costs, and find ways to expand.** "It took me a year and a half to figure out the right business model," Nyakarundi said. "And that came about with some of the failure we had in Uganda, which was our first expansion market. It's extremely challenging. When you bring in a technology that is not existing or nascent, you have to educate the customer; you have to show value. During all that time, you still have to have money and funding to continue operations. You've got to stay focused on the vision."
- **View entrepreneurship as a battlefield.** "Every day, you're fighting, solving a problem," Nyakarundi said. "Every day, you're going to do something to achieve your goal. There's not a day you're not going to be fighting, and it's a very lonely road. I always say, 90% of whatever you want to do as an entrepreneur is a mental game. You have to be prepared mentally, and don't fall into this trap of what you see on TV or magazines or social media, where you just start the business, and then in six to eight months, you'll be making X amount of money. That's just nonsense."

- **Consider finding a co-founder.** "Don't go on this journey by yourself," Nyakarundi said. "Find a co-founder who is as passionate as you are, who brings more value, who brings some skill sets that you don't have to balance you out. It will shift the weight that you carry."

Additional Journeys

Nyakarundi is one success story, but there are others driving change in Africa's renewable energy sector as well. A few of them are listed below. I hope seeing what they've accomplished will help aspiring entrepreneurs take the leap and start their own enterprises.

- Nthabiseng Mosia was born in Ghana but raised in South Africa. Mosia attended Columbia University in New York to study for a master's in energy finance and policy so she could have an impact on the energy grid in Africa. Today, Nthabiseng is the co-founder and CCO of Easy Solar, a Sierra Leone-based solar solution company.[10,11,12]
- Olivia Nava and Sachi DeCou started Juabar, which provides solar-powered mobile phone charging kiosks in Tanzania.
- Ugwem Eneyo co-founded SHYFT Power Solutions with Cole Stites-Clayton after experiencing power cuts and fragmented energy grids growing up and working as an ExxonMobil Environmental and Regulatory Advisor to gain experience.[13,14,15]
- Dave Lello, co-founder of Ekasi Energy in South Africa, started his career selling IBM PCs in the 1980s.

Ekasi produces home appliances designed to utilize solar power for electricity and wood pellets for cooking/heating. The idea is to provide a single appliance energy solution for urban dwellers.[16,17]

- Hannah Kabir studied renewable energy in Europe before starting Creeds Energy and working to encourage girls to embrace STEM careers so that all of Africa can have access to the basic necessity of electricity.[18,19,20]
- Habiba Ali set up an NGO to accelerate the adoption of renewable energy after her husband bought her a solar cooker from Germany. From there, she learned about the health risks of coal-powered appliances, and she began selling solar lanterns to address these concerns.[21,22,23]
- Derrick Hosea Opio founded One Lamp to end the reliance on dangerous and expensive energy sources such as kerosene with an on-demand energy service that provides safe, reliable energy to more than 5,000 people.[24,25]
- Lois Gicheru attended Strathmore University in Kenya before founding two energy companies and winning the Mandela Washington Fellowship for Young African Leaders.[26,27]

Acknowledging Our Next Steps

Internationally, governments and companies recognize that there are tremendous business opportunities in Africa's energy industry.

Entrepreneurship is one important strategy for keeping good jobs here, in Africa, where we can set up the infrastructure and green factories we'll need to really, really be able to do something. We can use Western technology in Africa. We can create new tech, too. We can take corrective measures to make African entrepreneurs and businesses competitive.

And, most importantly: We must do it for ourselves. We cannot wait for the West to do it for us.

Chapter 17

ENERGY DEVELOPMENT CAN HELP SOLVE AFRICA'S "MISSING JOBS" PROBLEM

According to the Brookings Institution, Africa has a jobs problem. For decades, Africa has been unable to provide its growing workforce with "decent, productive, and secure" employment, Brookings said—and things are only getting worse. Their 2020 research report pointed out how the scope of this problem was spreading, since jobs for young people are particularly scarce, and the continent's growing population is the youngest in the world.

There have been considerable efforts to expand training and even to prepare a young generation of entrepreneurs. But these initiatives and others like them have proven inadequate and sometimes even counterproductive. They've raised the skill level of many Africans, but they haven't created enough new jobs for African workers to fill.

Brookings noted the disappointing consequences of these programs in an article that accompanied the paper. "Churning out better-qualified and more motivated job seekers has not created more jobs—and instead has drawn funding and attention away from strategies that could create more jobs," it said. "The problem needs to be recognized for what it really is: a crisis of missing jobs."[1]

The development of Africa's energy resources—fossil fuels and renewables alike—can help fix this problem. It can create jobs for the large (and growing) number of Africans who want to work but can't find steady or well-paying employment.

This can be accomplished both directly and indirectly—by creating jobs in the energy sector itself, by developing jobs in related sectors, and by giving African countries the ability to produce enough electricity to support other types of businesses.

But it all hinges on one thing: investment in hydrocarbons and renewable energy.

Investment from the West is drying up in the case of hydrocarbons, due to blanket bans on funding fossil fuel projects, and investment is insufficient in the case of building solar and wind power infrastructure in developing nations. In fact, Western countries have put little to no money into Africa's renewables space.

Lack of investment in fossil fuels means Africans have to leave petroleum resources, our continent's riches, in the ground—or be branded an enemy to the environment. The result? Jobs that might otherwise be available to young Africans are like dust on a breezy day—infinitesimal, scattered, and increasingly out of reach.

While environmental causes are a major focus in the West, lawmakers in Africa's developing countries are more concerned with living wages and supplying basic necessities to the continent's growing population.

Which means African governments have to do more to support all energy projects that will increase electrification and put more Africans to work. That means being realistic about the barriers our own leaders put up to discourage energy development in Africa, corruption, and red tape among them.

Electricity: The Main Reason for Developing Energy Resources

Fossil fuels and renewable energy are different in many ways, but they serve the same purpose: generating electricity and jobs.

The development process for hydrocarbons—crude oil, gas condensate, natural gas—is extractive. It almost always involves using loud, heavy, and obtrusive machinery to drill a borehole that allows materials trapped below the surface to flow up and out. As such, it entails a certain amount of deliberate destruction and disturbance. It can't be done without ripping up soil, bedrock, and other sediments and then moving them out of the way. It necessitates the construction of roads and other infrastructure. Of course, governments can (and should) regulate the process, and responsible companies can (and should) try to minimize their impact on drilling sites and surrounding areas. But development simply isn't possible without reshaping the landscape.

By contrast, renewable energy projects are often less disruptive. They tap into phenomena that are already happening naturally—water flowing down a riverbed, wind blowing over the terrain, sunlight radiating from the sky. None of these things have to be pulled out of the ground; they're already out in the open. They tend to involve less drilling, less digging, and less extraction, so they are less destructive, even when they have to be tied into infrastructure networks just as oil and gas fields do.

Despite these differences, hydrocarbons and renewables serve the same end. They can both generate electricity—and as I've mentioned elsewhere in this book, electricity will be necessary for industrialization, economic diversification, and the improvement of living standards in Africa.

Industrialization and economic diversification create jobs. And the way to promote job creation is to support the development of energy sources with an eye to increasing the production of electricity for domestic consumption. Without this electricity, Africa will have a much harder time sustaining the businesses and industrial facilities necessary to develop other sectors of the economy and generate jobs over the long haul.

All of the Above, for as Many Jobs as Possible

Accordingly, African governments should support all projects that promise to increase domestic electricity supplies—and I do mean *all projects*. Not just oil, not just gas, and not just renewables, but all of the above! African countries need electricity in order to create and sustain more jobs and better standards of living on a long-term basis. As such, they shouldn't rule any options out.

In fact, the more options they employ, the more chances they will have to create jobs—not just indirectly, through the provision of fuel that can produce the electricity needed to sustain industrialization and activity in other sectors of the economy, but also directly. If a hydrocarbon-rich country such as Senegal opts to expand its wind power capacity, in addition to developing offshore gas fields, it will need to build wind farms as well as gas pipelines. As a result, corporations will need to hire more laborers and a larger number of local service providers overall.

Of course, many of these directly created jobs will be temporary. If a multinational green energy consortium hires local workers to prepare plots of land for the erection of wind turbines, it may then cut staff levels once all the equipment is in place. Or if an international oil company (IOC) works with two local service providers to secure enough welders to meet its deadline for completion of work on a new tank farm, it may then decide not to renew one provider's contract once they are back on schedule.

Even so, these temporary jobs offer some short-term benefits. And in the case of service providers, they also offer the chance to gain experience and expand capacity over the long term—to build larger and more impressive portfolios, as it were. To go back to the previous example, service providers who have worked with an IOC to complete a job on time will be able to demonstrate that their welders are capable of working quickly under pressure—and they can offer their services to other corporations, even if their contract with the IOC is not renewed. They can look for similar work in other sectors, or they can bid for additional contracts with other IOCs, perhaps even in neighboring countries.

If they succeed, they will need to hire more people to cover the additional assignments they accept. They'll need to create more jobs. If they do that, they'll be able to raise their profile even higher—and gain the attention of even more potential customers.

Policy Approaches for Encouraging a Virtuous Cycle of Job Creation

I've said before that part of the solution to Africa's energy poverty is to adopt a pro-entrepreneur stance. And by that, I mean that I'd like to see African governments adopt policies that discourage corruption, encourage transparency, and remove red-tape barriers to entry for new companies. These steps can encourage innovation and creative thinking, and they should also remove some of the obstacles faced by businesspeople who have good ideas but not many local connections.

Local content laws can be another part of this, and I discuss them more extensively in other parts of this book. But I also believe that *the way* countries approach local content laws is just as important as the laws themselves.

And that brings me to the case of Guyana. I want to talk about the approach this South American country is taking with respect to supporting local companies to expand and flourish as foreign investors rush in to explore and develop newfound oil reserves.

Guyana's New Frontier

Less than two decades ago, the most pressing question about Guyana, a former British colony on the northeastern coast of South America, was whether the country had the second smallest or the third smallest economy in the Western Hemisphere.[23] The country was small, poor, and obscure. Its population was less than one million, its government was highly indebted, and its natural resource endowment was seen as too small to attract much interest from investors.

Things began to change in May 2015, when U.S.-based ExxonMobil revealed that it had found oil in the first exploration well drilled at Stabroek, a deepwater block about 193 kilometers (120 miles) from shore.[4] This well, known as Liza-1, was just the beginning. The second exploration well drilled at the block—Liza-2, finished in June 2016—encountered deposits that were so large that ExxonMobil began describing Stabroek as a "world-class" asset, saying it might hold as much as 1.4 billion barrels of oil equivalent (boe) in recoverable resources.[5]

Liza was followed by more than 20 additional discoveries—the most recent of which, Whiptail-1 and -2, were made in July 2021.[6] There was so much oil that one of ExxonMobil's non-operating partners, the U.S. independent Hess Corp., raised its estimate of Stabroek's total recoverable resources to nine billion boe in October 2020.[7]

By the time Hess took this step, Guyana had officially entered the ranks of the world's oil producers. ExxonMobil had already launched production at Liza-1 in December 2019 and was talking about bringing the Liza-2 and Payara fields onstream in 2022 and 2024.[8] (Later, in April 2021, it named Yellowtail as its fourth development target and made plans to begin extracting oil in 2025.[9]) Meanwhile, other IOCs were drilling wells in the offshore zone in the hope of scoring similar successes, and foreign economists were realizing that Guyana had become the world's fastest-growing economy.[10]

Local officials were ecstatic—but also cautious. They fretted about the possibility that Guyana might succumb to "Dutch disease"—the apparent causal relationship between the increase in the economic development of a specific sector and a decline in other sectors—and stressed the need to establish institutions such as a sovereign wealth fund to manage and sustain oil revenues over the long term.[11] They also emphasized that they wanted oil projects to benefit ordinary citizens, not just the wealthy and the well-connected. To this end, they examined various options for promoting the creation of jobs and the development of valuable skills.

And the tool they chose to achieve this aim was local content policy.

Mark Bynoe, the former head of Guyana's Department of Energy, explained this choice in an address to the Guyana Manufacturing and Services Association in December 2019. He stated that the government had drafted a local content policy that could "ensure the highest possible rate of domestic employment, strengthen the local economy, and help the Guyanese economy to become less reliant on foreign capital and expertise."[12]

Bynoe was speaking at a time when Guyana's government was led by President David Granger, the head of the right-leaning APNU-AFC party. However, local content does not appear to be a partisan issue. Guyana's current President Irfaan Ali, who succeeded Granger in 2020 after a hotly contested election, has also been keen to develop a strong body of policy and law on this front—despite his position as the head of the left-leaning PPP/C party. His administration has continued discussions on the draft local content policy finalized by Granger's team in February 2020, while also speaking out about the importance of fostering the development of companies that can create jobs for Guyana's people.

For example, Vice President Bharrat Jagdeo noted in August 2020 that Ali was determined to make sure that oil projects opened up new jobs and new opportunities for Guyana's citizens. "We have made it clear in the meetings with ExxonMobil that we want them to do well, but Guyanese must share this prosperity. It is not sustainable otherwise, and we will insist that happens," he said in August 2020.[13]

Balancing Local Providers' Capacity with Investors' Needs

Why am I talking about Guyana's embrace of local content laws after saying explicitly that local content laws aren't the only necessary element in job creation?

Because I want to set the stage for another side of the story. I want to describe the backdrop against which Guyanese professionals and government representatives have been talking about how to make local content into an engine of job creation.

What they've been saying seems simple enough, that local businesses seeking contracts from IOCs ought to consider the needs of those IOCs as they work to build up their capacity.

In March 2021, for example, Floyd Haynes, a Guyanese-born accountant and management consultant who serves as a member of the government's Local Content Panel, said during an online discussion[14] that local companies shouldn't just expect contracts to come their way. Instead, he said, Guyanese providers must work to ensure that they are offering services that IOCs need—and that there is a balance between local capacity and local content. "We cannot just insist that [ExxonMobil] hires Guyanese [workers] and Guyanese companies without ensuring that these companies are adequately trained and prepared ... [If] we do that and those individuals are not properly trained, it increases the cost of doing business. And that cost rebounds to us as part of cost recovery, so the Guyanese taxpayer ends up paying for any sort of inefficiency," he asserted.

Haynes's sentiments were echoed by Vice President Bharrat Jagdeo, who was participating in the same online discussion of the government's efforts to finalize its local content policy and draft supporting legislation. He also stressed the importance of striking a balance between capabilities and content, saying: "We don't want to put something on paper that's totally unrealistic."

Staying in the Loop: Keep Local Companies Informed

It seems that the Guyanese government already has an idea of how to match Guyanese providers' capabilities with IOCs' needs while also leaving room for growth.

President Ali outlined that idea in February 2021, during stakeholder consultations on the country's local content policy. He said his administration wanted to require investors to inform the local business community of their long-term requirements so that Guyanese companies could develop their capacities in a focused and targeted way. Making this information available will allow local firms to become more competitive in a way that goes beyond local content targets or requirements, he declared.[15]

"Whilst the government will set targets, I think the operators [ExxonMobil and other IOCs] ... must be brought to a forum in which they say, 'This is our requirement over the next 20 years. These are the areas that we want, [where] we will have services that we will need,'" he said. He added that he was certain that if opportunities existed and Guyanese entrepreneurs were aware of these opportunities, they would build and expand capacity to meet opportunities.

Ali's statement wasn't exactly surprising, in that it echoed the language contained in the contemporary draft version of Guyana's local content policy. As the OilNOW.gy news site noted in February 2021, the document points out that Guyanese firms interested in working with IOCs needed to know what type, number, and level of goods and services might be required—and for what amount of time they might be required. "Only then can Guyanese institutions design and deliver training and development programs efficiently and effectively, or can businesses invest in equipment, materials, facilities, technology, etc., or parents invest with a sense of comfort in their children's higher education," the draft policy states.

Knowing Which Niches to Fill

I'm not going to pretend that I've given you a full and complete picture of what's happening in Guyana's oil industry. This isn't the time or the place for such an endeavor.

But I must say that I think Guyana is on to something here. And it's not because it's focusing so intently on local content policy and legislation.

Guyana's government is acknowledging that governments need to do more than set local content goals and then expect IOCs to follow them—or risk the consequences.

Officials in Georgetown, Guyana's capital, understand that there needs to be a balance between local capabilities and local content.

The Ali administration is aware that Guyanese companies would benefit from knowing which niches they can try to fill over the long haul. It's also because the administration is creating a framework in which IOCs can make their needs known and the local business community can grow to meet those needs.

I want to see Africa do something similar. I believe that every African country contemplating investment in energy—whether in oil and gas development or in conventional or renewable power generation—can learn from Guyana's example. I hope that African governments will approach local content not just as a matter of setting goals and monitoring compliance, but as a matter of matching local capacities to investors' needs and leaving room for growth.

That approach will help African service providers grow in a way that creates jobs, preserves jobs, and goes on to create more jobs.

Chapter 18

DIVERSIFYING ENERGY AND DIVERSIFYING ECONOMIES

Communism is Soviet power plus the electrification of the whole country.

—Vladimir Ilyich Lenin[1]

IN CASE YOU haven't figured it out by now, I'm a big believer in markets and free enterprise. I believe with all my heart that the best way for economies to grow and social conditions to improve is to ensure that governments don't stand in the way of business or try to dictate the course that companies must follow. (Regulations are necessary, of course, but they shouldn't prevent markets from working or companies from functioning and growing.)

So why, then, am I starting this chapter with a quote from one of the world's most famous communists?

Because I believe he had a point, in a roundabout way.

Let me explain.

Speeding Up the Process

After Lenin and his Bolshevik faction seized control of Russia in 1917, they found themselves facing a monumental task: providing the world with concrete proof of Karl Marx's ideas about how to move past the exploitation of industrial capitalism and on to socialism and, eventually, communism. Yet how could they possibly accomplish this ambitious goal when they had inherited a country that still exhibited vestiges of feudalism? How could they move the country beyond capitalism when it hadn't even industrialized sufficiently to establish a modern standard of living? After all, it was still a poor and backward place, where wooden plows had arguably remained more common than sophisticated manufacturing plants, even in the early 20th century.[23]

Fortunately, Lenin had already added his own twist to Marx's categorical assertions about the path of history by positing that Russia didn't need full industrialization to start moving into the next phase of socialism. And since the Russian revolution played out in a way that convinced him that his theories were correct, he was confident that all of his subsequent policy decisions—including the call for electrification that I cited above—would also be vindicated.

His confidence was very much on display in his explanation of why he wanted the Soviet Union to build as many new power plants as quickly as possible. "There can be no question of rehabilitating the national economy or of communism unless Russia is put on a different and a higher technical basis than that which has existed up to now," he said during a speech in 1920. "Communism is Soviet power plus the electrification of the whole country, since industry cannot be developed without electrification."[4]

I'm focusing on those remarks in particular because I think they contain a grain of truth:

Electricity is critically important for economic diversification.

Lenin had a point.

And, yes, I am arguing that economic diversification is what Lenin was aiming for. He might not have put it in those exact terms, but think about it: What did he want power plants to do? He wanted them to supply enough electricity to make the Soviet Union capable of supporting many more factories and many more types of industry than it did (or could) at the time. He understood that factories and industries were necessary to the type of economic development that could support modernization and higher standards of living. He understood that development would be easier to sustain if it did not hinge on one single type of economic activity.

That, then, is where I agree with Lenin.

I want to see African economies develop in a way that can support modernization and higher standards of living. Development will be easier to sustain over the long haul if it doesn't focus on a single sector of the economy.

I know that economic development involves industrialization and diversification.

And I know that industrialization and diversification are impossible without electricity.

A Roadmap for Using Every Available Resource

Fortunately, Africa has plenty of what it takes to produce electricity. The continent has abundant reserves of hydrocarbons—at least 125.3 billion barrels of crude oil and 14.41 trillion cubic meters (509 trillion cubic feet) of natural gas, according to *GEO ExPro* magazine.[5] It also has the resources needed to support renewable energy projects—enough sunlight to support 1,000 GW of solar generation, enough running water to support 350 GW of hydropower generation, enough wind to support turbines that can generate 110 GW of power, and enough geothermal energy to support 15 GW of capacity, according to an estimate that the African Development Bank (AfDB) made in 2017.[6]

We ought to be utilizing these resources—all of them—to the fullest extent possible, in the most sustainable way possible. Here's my vision for what that might look like, in brief:

- **Start with gas**: I can't emphasize this point enough. The primary goal should be to boost African electricity production dramatically and rapidly. If so, gas—and I'm talking about both associated gas, which can be separated out during oil production operations, and natural gas—is the best way to start. Africa has a lot of gas, and we already know how to use it to generate electricity on a large scale. Even better, we know that it burns more cleanly than fuels made from crude oil. Let's use as much of it as we can—and let's use it to support domestic electrification projects, as well as generating export revenue.
- **Explore options for oil:** Africa shouldn't ignore oil just because petroleum products—gasoline, diesel, jet fuel, kerosene, residual fuel oil (RFO), etc.—generate more harmful emissions. Instead, it should focus on gas for power generation wherever possible and take advantage of a larger share of oil production to turn out value-added industrial materials such as petrochemicals and lubricants, which command higher prices than traditional refined fuels. This shift could support Africa's oil-producing states in generating more export revenue and give them an opportunity to establish themselves as suppliers of vital commodities to African industrial organizations.
- **Phase in renewables:** Even if gas is the starting point for boosting Africa's electricity production, it's not the only runner in the race. The ultimate goal is (and should be) to provide the continent with the optimum energy mix—the combination of fuels and inputs that can produce the largest amount of

electricity in the cheapest, fastest, most efficient, and most sustainable fashion. Right now, that mix is heavy on gas, precisely because renewables don't pack as much of a punch. (That is, they don't yield as much energy for the money as fossil fuels.) But this is not likely to remain true in perpetuity. Solar and wind technologies are improving—and as they do, they should take up a larger share of the energy mix.

- **Use energy to support diversification**: Produce more electricity to support other sectors of the economy that can create jobs and expand the economy—agriculture, manufacturing, tourism, and so on. Meanwhile, look for ways to expand the energy value chain. Use both gas and oil to produce petrochemicals and other high-value industrial materials. As renewable generating capacity grows, use solar and wind plants to produce green hydrogen, as well as electricity. Seize every opportunity to acquire new technologies and skills along the way, and work with partners who can help train Africans to use them.

Energy Development Inevitably Supports Diversification

Let me put it another way: Electrification will open the path to economic diversification, and the process of developing the resources needed to support electrification can act as a springboard toward economic diversification.

"Dutch disease" doesn't have to be the inevitable result of oil and gas extraction—provided that African producers embrace hydrocarbons not just for their own sake, but also for their ability to serve as a bridge to other types of economic activity. In practical terms, what this means is looking as far down the value chain as possible—and examining every link in that chain. It means using hydrocarbons to support refining, petrochemical, and gas-to-power

(GTP) projects, all of which can then lend momentum to other sectors of the economy.

In the case of natural gas, for example, producers can invest in fertilizer plants and then incorporate their factories' output to support the agricultural sector. They can also build petrochemical units capable of supplying plastics, chemicals, and other materials with high added value. Additionally, they can establish links to GTP initiatives, which support the industrial sector by providing manufacturers with electricity so they can remain in production and avoid outages.

Alternatively, they can arrange to supply gas directly to heavy manufacturing concerns, as companies working in the Eagle Ford shale basin in South Texas have done for factories in northern Mexico.[7] They could even produce LNG for powerships—that is, seagoing vessels that can travel to the destinations where electricity is needed most—as Senegal has arranged to do under a deal with Turkey's Karpowership that is being financed by Mauritius Commercial Bank.[8]

Likewise, as renewable energy technologies develop, they will also generate their own downstream value chains. They will, for example, give countries with high solar potential the chance to explore desalinization and green hydrogen projects, as well as electricity generation.

They'll also create supply chain opportunities since several African countries possess reserves of the rare earth elements (REEs) that are needed to manufacture wind turbines[9]—and, to a lesser extent, solar panels and batteries.[10]

I happen to think this could be a very good thing in the long term. African countries should have the chance to provide raw materials for the renewable projects that will power their own communities. And if they find the right partners, they should have the chance to learn how to make those raw materials into finished products. Imagine what might happen if a Kenyan investor had a choice between importing everything it needed to build new turbines in the Turkana Wind

Corridor[11] and buying locally produced parts, including permanent magnets made with neodymium from Burundi.

Moving Down the Value Chain: The Example of Renergen

There are so many ways to tie energy production into other sectors of the economy.

Consider, for example, Renergen's Victoria project, which centers on the development of several gas fields in South Africa's Free State.[12] This company hopes to extract gas from fields near Welkom, Virginia, and Theunissen and process it at a liquefied natural gas (LNG) unit that Renergen began building at its existing compressed natural gas (CNG) plant in Virginia in November 2019.[13]

Renergen already has plans for its LNG, which is expected to have a higher profit margin than CNG. In June 2020, the South African company signed an agreement with the French major Total (now TotalEnergies) on future production from its gas liquefaction plant. Under that agreement, it will team up with a local subsidiary of TotalEnergies to make its LNG available for domestic purchase by long-haul truckers. Specifically, it will supply LNG from the first phase of its plant to TotalEnergies for distribution via that company's branded filling stations along the Johannesburg-Durban section of the N3 highway. The agreement with TotalEnergies has already helped it secure at least one major customer—Logico Logistics Group, a local niche market transport service provider, which will buy Renergen's LNG from TotalEnergies-branded stations under an agreement signed in October 2020.[14]

The partners hope to expand their operations in 2023 when the second stage of the LNG unit is due to begin production. At that point, Renergen will start sending part of its production from the facility's second stage to the French major for distribution in TotalEnergies retail outlets along all of South Africa's major highways, while the remaining volumes will be reserved for other buyers.

Renergen's plans may not sound terribly novel, given that they put natural gas to work in the road transport sector, which has been heavily dependent on refined versions of fossil fuels for more than a century. I'm going to argue, though, that they provide important (if indirect) support for economic diversification.

I base this assertion on the fact that the deal between Renergen and TotalEnergies will allow South Africa to employ domestic gas production to support the operation of critical infrastructure—namely, domestic roads.

Roads may seem like mundane things, but their value shouldn't be underestimated. South Africa has the continent's longest domestic road network, measuring nearly 750,000 kilometers. In turn, this network plays a key role in freight deliveries, handling 70-80% of all domestic goods shipments.[15,16] It generates more than USD1 billion worth of income each year[17] and accounts for at least 5% of total GDP.[18] It is indispensable with respect to moving goods out of large landlocked cities such as Johannesburg to the coast for export, and it offers overland connections to neighboring markets in Zimbabwe, Namibia, Mozambique, Botswana, and other countries.[19]

In other words, roads are critical for the trucks that keep South Africa's economy running. They accommodate the trucks that support the country's many different industries, some of which are very high-value indeed—mining, manufacturing, agriculture, and consumer goods.

In turn, all these trucks need fuel. South Africa does produce some of the petroleum products required to fill their tanks, but its production of diesel and other fuels has gone down as a result of refinery closures and other problems in 2020.[20] Renergen's LNG will help cover this gap without forcing the country to import more, and as such, it will support other sectors of the economy beyond the gas sector itself.

What's more, Renergen will do so while reducing emissions, since LNG generates less pollution than diesel fuel, traditionally the main fuel for road freight.

At the same time, though, this LNG project will also make a direct contribution to economic diversification by creating new possibilities for the trucking industry. Renergen has also joined forces with Henred Fruehauf, a manufacturer of trailers, to develop a zero-emissions solution for refrigerated road freight.[21] This solution uses the company's LNG, which must be stored in fuel tanks at a temperature of -162°C and then heated to an ambient temperature of approximately 15°C before combustion, to cool the cargo holds of long-haul trucks. It appears to be cost-efficient as well as less polluting, as it does not require large amounts of extra fuel or energy to operate. According to Renergen's web page, this partnership reduces the total amount of greenhouse gas (GHG) emitted by each truck-and-trailer combination by up to 96 tons per year, while also cutting total fuel costs by 23%.

Electricity Supports True Economic Diversification

Renergen's LNG project is designed to revolutionize fuel markets in South Africa, and its contribution to economic diversification will mostly be evident in that country. Electrification, by contrast, could effect change on a much larger scale, transforming economies across the African continent.

By that, I mean that if Africa utilizes its energy resources in a strategic way, starting with gas and then shifting gradually to renewable energy as technologies improve, it will be in a better position to establish and maintain the industries, infrastructure, and institutions that will allow it to thrive. It will have more (and better) opportunities to work toward true economic diversification, in which economies don't succeed or fail on the basis of a single crop or commodity, and people have more economic choices and opportunities.

Certainly, electricity makes industrial development easier. It can power new plants that produce new goods and sustain new jobs, such as the massive oil refinery and petrochemical plant that the Dangote group is building in Nigeria's Lekki Free Economic Zone (FEZ).[22] It can power innovation hubs that generate new technological solutions, such as the facilities that have established Kenya as "the heart of East Africa's technology ecosystem," in the words of the GMS Association (GSMA).[23] It powers machines and lights up job sites for construction workers who build new factories. It can power the heavy equipment needed to dig new shafts at places like the Ewoyaa lithium mine in Ghana[24] and run the ventilation systems that will bring air and light to the miners working in those shafts. It can power the computers, cyber-physical monitoring devices, and control systems that have given companies like RCL Foods, a South African sugar producer, a sharper technological edge.[25]

But electrification doesn't just support industries and the industrialization process. It also powers the infrastructure networks, schools, training facilities, stores, and transportation routes that benefit factory workers, students, and their families. And as such, it gives a larger number of people more opportunities to improve their lives.

With more electricity, African countries have a better chance of building up their industrial base.

With more electricity, African countries can shore up the sectors of the economy that support industry.

With more electricity, African countries can build better roads, transportation networks, and other types of infrastructure.

With more electricity, African countries can offer their people more education and training.

So perhaps I should put my own twist on Lenin's maxim and assert: *Economic development is the electrification of the whole African continent.*

A Vision of Senegal's Future

But what would this look like in practice? What might it mean for an African country to use every possible option to increase electricity production in order to support economic diversification?

Let's try to imagine it.

We can start by assembling a small collection of facts about a single African country: Senegal.

- Senegal lies along the West Coast of Africa, a region where the Atlantic trade winds have the potential to generate thousands of terawatt-hours of electricity each year, provided that enough wind turbines can be built.[26]
- Senegal possesses substantial gas reserves, both offshore in fields such as Yakaar and Teranga, which BP (U.K.) is developing in cooperation with Kosmos Energy (U.S.)[27] and the national oil company Petrosen, and onshore in fields such as Gadiaga, where Fortesa International (U.S.) is already extracting around 84,960 cubic meters (three million cubic feet) per day of gas.[28]
- Senegal plans to use some of its future gas production to fill a pipeline that will serve power plants, thereby bringing the country's generation capacity up to 2.5 GW by 2030.[29] The U.S. Trade and Development Agency (TDA) has already agreed to fund a study for this project.
- Senegal will also provide gas to fuel a powership that will generate low-carbon electricity for the domestic market.[30]
- Senegal's main economic drivers are primary agriculture and the service sector,[31] both of which would benefit from improvements in electricity supplies.
- Senegal has the potential to diversify its economy by investing in rice production, specialized agriculture, tourism, and other endeavors.

So let's assume that Senegal continues the work it is already doing with BP, Kosmos Energy, Fortesa, and other international oil companies (IOCs) to develop onshore and offshore gas reserves, and with USTDA and other organizations to build pipelines. Let's assume it keeps working with investors to construct more gas-fired power plants similar to the one GE (U.S.) and Calik Enerji (Turkey) are building in Cap des Biches.[32] Let's assume its deal with Karpowership is implemented according to plan. And let's assume that all these initiatives do a good enough job of supporting domestic electrification that Dakar can turn to setting up wind farms that are even larger than the 158-MW facility launched at Taiba N'Diaye by Lekela (U.K.) in February 2020.[33]

Then let's imagine that all these efforts work well enough over time that Senegal's government achieves its stated goal of bringing the country's generating capacity up to 2.5 GW by 2030, nearly triple the 2020 figure of 864 MW,[34] and achieving nationwide electrification.

Let's imagine, in turn, that all this new capacity is used to strengthen the service sector, expand the range of agricultural production, and develop the tourist trade. Let's imagine that the expansion of these sectors fuels growth in other industries, such as food processing and telecommunications. Let's imagine that new educational institutions and training centers emerge so that Senegal's people have opportunities to gain more skills, more knowledge, and better jobs. Let's also assume that these developments are accompanied by efforts to expand transmission and distribution networks to every part of the country so that the gains aren't confined to the cities.

Doesn't that sound like the kind of economic diversification Africa needs? I think it does.

And that's why I believe the continent should work to develop all of its energy resources to the fullest extent possible—not just fossil fuels and not just renewables, but both, and in the sequence that makes the most sense. African economies need all the support electricity can provide. And if we want electricity, we have to do what it takes to produce it—even if that means quoting Lenin and unnerving environmental activists by choosing gas-fired generation as our starting point.

Chapter 19

AI, ANALYTICS, AND INNOVATION: HOW AFRICA CAN MAKE THE LEAP TO NEW TECHNOLOGIES

IT'S NO WONDER that economics is so often described as the "dismal science." Textbooks and academic works often reduce economics to a set of formulas, theories, and assumptions defined in terms that are grey, dusty, and abstract, seemingly removed from the real-life phenomena they describe.

Every so often, though, we get the chance to see a concept from economics play out in real life. I'm talking about the concept of leapfrogging,[1] and I've seen it happen in the realm of telecommunications in Africa. Let me explain: Africa began the 21st century at a stark disadvantage to developed countries on this front. As of 2000, fewer than 3% of its rural settlements had access to fixed phone lines, and conditions weren't much better in urban areas. The entire continent had fewer fixed telephone lines than the island of Manhattan![2] Mobile telephones were still rather rare, and internet connections were few and far between. As a result, far too many Africans faced obstacles when they tried to expand their entrepreneurial and personal horizons beyond their own immediate surroundings.

By 2019, though, there were telephones everywhere. Specifically, there were hundreds of millions of Africans in possession of mobile phones, and the number was on track to top 1 billion, more than 83% of the total projected population of 1.2 billion, by 2022.

So how did this change happen, exactly? Did Africa follow the familiar pattern of Western nations, where the telecommunications industry began with a few fixed lines before expanding to encompass nationwide and international service, mobile communications, the internet, and wireless connections? Did it start from scratch and then pass through every known stage of evolution?

No, it did not. Instead, the continent leapfrogged past the stage of establishing large fixed-line networks with analog controls and moved directly to setting up digitally controlled wireless networks—and not just any wireless networks, but the more advanced ones. As Vincent Kaabunga, the chairman of the Ad Hoc Committee on Activities in Africa of the Institute of Electrical and Electronics Engineers (IEEE), noted in 2019, "Some places will go straight to 5G."[3]

But Africans have been quick to grasp the possibilities that these new technologies offer, especially smartphones with internet connections. They have, for example, turned the continent into a leading developer of innovative financial technologies, also known as fintech or "mobile money."[4] This drive to innovate was already evident in 2005, when Vodafone and Safaricom launched M-Pesa, the mobile payment service that quickly spread out from Kenya to become the largest mobile money service in the developing world.[5]

That drive is still evident now, and it led Obi Ozor and Ifo Oyedele, two Nigerian entrepreneurs, to try their hand at rolling out a new business model for freight and logistics in Africa. Ozor and Oyodele created Kobo360, an Uber-like platform that matches truck owners and drivers with available capacity to shippers in various locations, with the goal of helping businesses move products to market more quickly while also helping truck drivers and owners increase their earnings by never traveling empty.

Since its launch in 2017, Kobo360 has expanded its reach beyond Nigeria into Ghana, Kenya, Togo, Uganda, Burkino Faso, and Côte d'Ivoire. As of late spring 2022, it had already created 150,000 jobs and had been used by tens of thousands of drivers transporting more than nine billion kilograms of freight for a wide range of shippers, including big-name customers such as Maersk and Unilever.[6]

African Solutions for African Challenges

In my opinion, these achievements are impressive on an objective level. But that's not why I'm singling them out. I'm drawing attention to them because of the circumstances under which they occurred. They were a product of leapfrogging.

African countries and African businesses were able to overcome the disadvantage of establishing telecommunications networks late in the game by adopting the best new technologies available as one of their first steps. As a result, they were able to minimize the time, money, and effort invested into older solutions such as wired telephone lines and analog switchboards. They didn't get tripped up by sunk-cost considerations. Instead, they just skipped the intermediate steps and traveled a shorter path toward widespread wireless communications.

Even better, they created African solutions for African challenges. M-Pesa didn't succeed simply because it was innovative and useful; it also offered customers who had no reliable access to banks new ways to send and receive money without shouldering the risks of traveling over long distances with cash. As Daniel Rohio, an M-Pesa customer in Kenya, told the Voice of America in 2009: "If I wanted to send money to my mum at a particular time or my grandmother, she cannot receive that money and would have to wait for maybe at least one day or two days. So I have found M-Pesa a little bit easier to send money to them right now."[7]

Likewise, Kobo360 drew attention from the likes of Goldman Sachs, which helped it raise USD20 million in funding in August 2019, partially because it was able to apply Uber's on-demand formula to road freight. It also won the loyalty of its users by making their journeys safer, as well as more efficient and profitable. That is, Kobo360 keeps truckers informed about weather and traffic conditions while also allowing them to avoid routes where armed bandits might target them. This steady flow of information allows Kobo360 to reduce the time needed to complete a 1,000-kilometer journey from a week to just three days, Ozor told CNN in 2020.[8]

Why am I spending time on these stories? What does all this have to do with energy? I think Africa has the opportunity to accomplish a similar feat with respect to technology and the petroleum industry. Let me explain a little further.

AI and Analytics: A Good Fit for the Energy Sector

Within the last two decades, the world has entered a new stage of digitalization. (Some have gone so far as to call it the Fourth Industrial Revolution.) In this new stage, technology isn't just a means to an end; it can also be a part of the decision-making process.

The world is now in a position to supplement human intuition, experience, and inspiration with tools such as artificial intelligence (AI) and analytics. Those tools are capable of processing staggeringly large sets of digitized information. (There's a reason why it's called Big Data!)

These sets may include data points collected in real-time by Industrial Internet of Things (IIoT) devices—sensors and monitors connected to the internet—and/or data points collected in the past from other settings. But the tools don't just process the data. They also look for patterns within the whole body of information, and seek ways to apply those patterns to real-world situations.

As a result, businesses with access to these tools have more ways to make their operations more cost-effective, more efficient, and more responsive to customer preferences and market fluctuations, as well as safer and less subject to breakdowns and outages. They can learn whether employee accidents follow any particular patterns and identify ways to eliminate hazards. They can figure out what conditions make equipment failure more likely and design predictive and preventative maintenance routines to reduce downtime. They can reveal when and where raw materials, equipment, and labor are most productive, thereby reducing waste and increasing efficiency. They can predict the likely impact of specific seasonal factors or changes in environmental conditions, thereby gaining their users time to stock up on raw materials, transfer products in or out of inventory, and ride out market fluctuations. They can identify shifts in customer preferences and tailor production schedules accordingly.

In short, these sophisticated tools have the potential to help businesses do more and perform better while spending less.

What's more, AI and analytics aren't restricted to any particular arena of business. They're already in use in the energy sector. In fact, geophysical service providers were among the first companies to recognize the value of analytics and Big Data as they conducted large-scale seismic imaging of oil and gas reservoirs.[9]

They're also present at every link of the value chain.

With respect to electricity, Uniper, a German energy supply company, says AI can help conventional thermal power plants (TPPs) reduce pollution by achieving fuel-to-air ratios that can maintain optimal combustion temperatures.[10] Tilt Renewables, an Australian wind farm operator, has used AI software to develop and execute a "precision trading" strategy that keeps its electricity competitive in price.[11] Azuri Technologies, a British start-up company, uses machine learning (a type of AI) to manage solar panels and the batteries they charge in ways that reflect customer usage data.[12]

On the oil and gas front, Orpak Systems, an Israeli software firm, says its analytics solution can allow petroleum product retailers to build customer loyalty and manage fuel stocks.[13] Repsol, a Spanish oil company, has teamed up with Google Cloud to utilize AI and Big Data solutions to optimize energy consumption, operational performance, and competitiveness at its refinery in Tarragona.[14] MISTRAS, a New Jersey-based company, designs predictive analytics solutions that allow pipeline operators to avoid disruptive and damaging incidents.[15] QRI, based in Texas, uses AI to power the SpeedWise reservoir management platform, which it markets to upstream oil and gas companies engaged in complex tasks such as integrated well planning.[16] Teradata, a California software firm, has designed BitBox to serve as a forensic analytics solution for drilling contractors who want to optimize the use of their drill bits.[17]

It may already be clear where I'm going with these examples, but I'll spell it out anyway.

Every company I just mentioned is contributing to the development of technologies with the potential to improve performance in the energy sector. Therefore, every African country and every African company involved in energy projects ought to embrace these technologies.

Yes, I mean *every single* country and *every single* company. Yes, even if they were originally meant to benefit non-African users working in non-African environments. Yes, even if they were designed to operate in places where energy markets and infrastructure are, on average, more developed than they are in Africa. Yes, absolutely!

Why?

Because the existence of these technologies and tools gives Africa the chance to engage in another bout of leapfrogging. If African stakeholders follow this path, they'll have opportunities to save time and money. They'll be able to do more than just develop their own hydrocarbon resources, expand their own electricity production, and support their own local electricity and fuel markets. They'll be able to adopt and integrate new technologies as they do all of these things. They'll be able to explore every option for making the African energy sector as efficient and productive as possible.

In short, I want Africa to take advantage of all the innovations and inventions that are available now, even if they weren't available in Africa first. So let's skip as many intermediate stages of development as we can. Let's set milestones for new technologies as we pursue our goals for energy.

Sticker Shock and Other Obstacles

I don't expect any of this to be easy.

For one thing, there will be significant upfront costs.[18] AI solutions and IoT devices cost money, as do the equipment, services, infrastructure, training, and personnel needed to support them.[19] As such, African countries and companies that make a point of including technology upgrades in their business plans and development strategies will face higher expenses. They'll also have to justify the extra costs when they negotiate with financial institutions and international oil companies (IOCs).

There may also be some logistical difficulties, in that much of the equipment and personnel involved in these new solutions will have to be imported or contracted from abroad.[20] African stakeholders and their partners will, therefore, need to make certain that they can deliver the right resources to the right locations. (They should also check beforehand to determine whether local customs regimes accommodate such deliveries.[21])

But I'm confident these challenges can be overcome. And the best way to do that is to think ahead.

If there are problems with the customs regime—say, tariffs that make the import of certain types of equipment prohibitively expensive—African businesses and governments should work together to change the rules (or decide on criteria for making exceptions).

If there are problems with logistics, African stakeholders should make a point of addressing the challenges before they make a formal decision to proceed with the investments. If, say, an African midstream operator is looking to establish a new network of smart pipelines in a location without reliable electricity supplies, it should extend the transmission grid as a component of its business plan and its budget, so that there are no surprises for the partners and financial institutions involved in the project.

And if there are problems with high-cost projections, African stakeholders need to build the case for spending that money. They should be ready to explain that leapfrogging can save money because it allows for skipping intermediate steps. They must be prepared to talk up the potential of Africa's internal market—to highlight the ways that efficiency can boost the profits of companies serving the continent's growing energy demand. They ought to have big-picture ideas about how to ensure that these new technologies benefit other sectors of the economy beyond energy.

What's more, African stakeholders should draw attention to the resources they already have in the technology arena—namely, their own people and their own experience in building new digital realms from scratch. They should be able to point to examples of African innovation and entrepreneurship such as Jobberman (www.jobberman.com), which has become Sub-Sahara's largest job placement website since its establishment in 2009,[22] and make a strong case for investing in more African innovators and entrepreneurs. They should think about how local technology hubs such as Nairobi's Silicon Savannah,[23] where hundreds of start-up companies have set up shop, might support IOCs and look for new opportunities in the energy sector.

Bringing Technology to Africa—and Africans

Stakeholders must consider how to bring technology to Africans as human beings who learn, work, and innovate, not just to Africa as a geographical entity.

By that, I mean they ought to look for ways to develop new tech talent in Africa and take advantage of the talent that already exists. They should pursue deals with outside partners that provide for training as well as technology transfers so that expat workers aren't the only ones capable of using these new tools. They must make connections between local tech experts and their foreign counterparts in order to support the development of new tech hubs and lay the groundwork for further cooperation.

Stakeholders ought to consider whether they can contribute to technology supply chains by providing equipment manufacturers with raw materials, such as minerals or metals, or low-cost labor for assembly and packaging.

Fortunately, there are already frameworks that support this—namely, local content regulations. Most oil- and gas-producing states in Africa have enacted laws that require IOCs to purchase, hire, or contract a minimum amount of local goods and services, and most of those local content laws make explicit provisions for technology transfers and training of local personnel. Ghana, for example, requires foreign contractors, subcontractors, licensees, and other allied enterprises to devise a program for transferring technologies to Ghanaian citizens or companies.[24] Angola, meanwhile, encourages IOCs to draw up annual training plans for local employees and to submit their plans to the Ministry of Petroleum ahead of implementation.[25]

As far as I can tell, though, these regulations don't specifically identify Fourth Industrial Revolution technologies, including but not limited to AI in all its varieties: IIoT, analytics, cloud computing, 3D printing, autonomous vehicles, and nanotechnology,[26] as priorities. I'd like to see that change. I'd like to see African governments make deliberate efforts to foster the development of advanced technological capabilities.

But I also believe African stakeholders shouldn't rely solely on local content laws. I believe the continent's tech innovators and entrepreneurs should put themselves forward. They should offer to help IOCs bridge technological gaps that have the potential to slow the pace of development. They should team up with local vendors to expand their range of services and improve the quality of services provided. They should join forces (and work with local chambers of commerce and other business groups) to arrange exhibitions where they can showcase their own capabilities to energy operators. They should actively seek funding for innovation hubs and training centers. They should be ready to show the world that Silicon Valley isn't the only source of creative solutions.

To a significant extent, Silicon Valley itself is already aware of Africa's potential—as demonstrated by Google's 2018 plan to build an AI research center in Accra, the capital of Ghana.[27] Moustapha Cisse, the head of the center, told Agence France-Presse in 2019 that he believed technology and training had the potential to change the continent for the better. "Africa has many challenges where the use of AI could be beneficial, sometimes even more than in other places ... We just need to ensure that the right education and opportunities are in place. That is why Google is sponsoring a lot of these young people for their degrees ... to help develop a new generation of AI developers," he said.[28]

So let's spread that awareness into the energy industry. Let's make IOCs aware that African sources can provide the technologies, skills, and drive needed to maximize efficiency and profitability.

But we must make sure that this heightened awareness doesn't lead to human rights violations—or to the exploitation of Africa's resources, and more importantly, its people.

Chapter 20

COBALT BLUES: CAN WE KEEP THE EV REVOLUTION FROM EXPLOITING AFRICA?

Any newcomer to the clogged and frenzied freeways of Los Angeles (L.A.), California, is probably impressed by the volume of traffic, but also by the variety of vehicles edging bumper-to-bumper for miles on end. Semi-trailers, pickup trucks, and VW buses that have seen better days jockey for space alongside Bentleys, Jaguars, and more Teslas and Nissan Leafs than you're likely to see anywhere else. After all, L.A. may be best known as the entertainment capital of the world, but it has also achieved a similar spotlight as the American city with the most electric vehicles (EV). Looking around L.A., there's no question the EV revolution has begun.

As EV drivers navigate the 101, it's unlikely they're giving much thought to the lithium-ion batteries that power their rides. As far as they're considered, they've made a climate-friendly choice. However, their "clean" decision could very well be tied directly to a dirty little secret more than 9,000 miles away in the Democratic Republic of Congo (DRC). The cobalt critical to their battery's function may have been picked by African miners working in hazardous conditions—or, worse yet, scavenged by a school-age child at an industrial site.

The DRC produces about 60% of the world's cobalt, much of it for Chinese companies where occupational safety takes a back seat to efficiency and profits. The southern part of the DRC is home to where most of the "artisanal," or informal, mining occurs—and where child labor abuse has been so severe that Amnesty International linked cobalt extraction to human rights violations.[1]

This energy story is not quite the earth-friendly news story anyone is hoping for. In fact, it seems more like just another page in the historical exploitation of Africa's natural resources and human capital. It doesn't have to be this way.

In addition to cobalt, Africa is rich in many of the other minerals that are among the key building blocks of EV batteries. With the International Energy Agency (IEA) saying there is a "looming mismatch between the world's strengthened climate ambitions and the availability of critical minerals that are essential to realizing those ambitions,"[2] we must find a way to leverage our mineral wealth and keep it from becoming the next blood diamonds. We must protect our people. We should be participating as a full and respected partner in the EV race.

Participation as equal partners will take a comprehensive effort by African leaders to get there. That includes applying the same strategies to our minerals industry that I've recommended for capitalizing on our oil and gas resources—including good deal-making that doesn't shortchange Africans; strong content policies that ensure employment, training, and business opportunities for locals; transparent revenue management by governments; and monetization programs with the potential to grow and diversify African economies.

For decades, we missed out on opportunities to reap the full benefits of our oil and gas industry. We must not make the same mistake with our mineral resources.

But before we can figure out a fix, we have to understand what's getting in the way.

Surging Demand

There's no question that from a tailpipe emissions standpoint, EVs are better for the environment than conventional internal combustion engine (ICE) vehicles, making them important foot soldiers in the march toward net-zero. In hopes of helping the world meet climate change goals (and capturing the climate-conscious buyer market), automakers everywhere have said they'd switch from petroleum to battery-electric vehicles (BEV) by the mid-2030s,[3] less than 15 years from now. By 2040, we could see 677 million zero-emission vehicles (ZEV) on the roads worldwide. That's a sizable increase over even the industry's own estimates; previously, BloombergNEF (BNEF) had predicted 495 million ZEVs by 2040.[4]

As every EV that replaces a conventional car, the expectation is that it will reduce total greenhouse gas emissions, or at least keep them in check. Obviously, a growing EV fleet will require more battery storage. In fact, the IEA estimates that by 2040, the world will need 10,000 GW-hours of batteries. That's an astonishing 50-fold increase over what's currently available.[5]

And there's more to EV batteries than cobalt: A single NMC532 lithium-ion car battery pack typically contains around 8kg of lithium, 35kg of nickel, 20kg of manganese, and 14kg of cobalt.[6] (Conventional cars, by contrast, require just one-sixth the mineral inputs of EVs.) Compared to 2020 levels, the world will have to produce 40 times more lithium and nickel and 20 times more copper, graphite, and cobalt just to keep up with growing EV battery demand.[7]

The increase in EV battery production should lead to a windfall for Africa, which has many of those minerals under its feet. In addition to the DRC's extensive cobalt reserves—3.5 million metric tons as of 2021, or nearly half of the world's 7.6 million metric ton total[8]—at least six other nations hold key battery minerals. South Africa, for example, is the world's largest producer of manganese, and has substantial nickel and platinum reserves; Zimbabwe and Namibia rank among the world's top 10 lithium producers; Mozambique and Tanzania produce graphite, and Zambia, copper.[9]

Sadly, though, the continent has not optimized the financial benefits that should accompany its resource wealth, largely because the focus has been on extraction and export of raw materials, rather than on the entire lithium-ion battery value chain. Beneficiation—that is, the treatment of raw materials to improve their economic value—refining, and recycling all take place outside Africa. Most of this work is done in China, which is seeking to dominate the global supply chain "from the metals in the ground to the batteries themselves, no matter where the vehicles are made,"[10] according to *The New York Times.*

China is off to a great start, by the way. Of the 136 lithium-ion battery plants in the pipeline to 2029, 101 are in China.[11] The country has 558 GWh of battery manufacturing capacity, more than anyone else, and by no small measure: The U.S., which comes in second, has a paltry (by comparison) 44 GWh.[12] China also controls more than 80% of the world's cobalt and graphite refining capacity,[13] meaning it can set prices.

It should surprise no one that China has looked to Africa to supply most of the raw materials it needs. In fact, it's safe to say China relies on Africa to keep its battery factories humming. In 2019, for example, graphite exports from Africa to China grew 170%. Almost 70% of the Congolese mining portfolio is under Chinese control. (Swiss Glencore, which runs two of the DRC's largest cobalt mines, is the most significant alternative to Chinese operators.)[14]

Certainly, this represents at least a dotted line relationship between Africa and a future world where there are fewer greenhouse gas (GHG) emissions.

Like most other things, though, there are tradeoffs for every benefit. In this case, environmental stewardship is coming at a societal cost for Africa.

"Using People Like Slaves"

Anywhere else, the situation in the DRC might be an anomaly, but in Africa the scenario is all too common. Although the country's untapped resource wealth is worth at least USD24 trillion, its citizens remain mired in poverty.[15]

It's easy to understand, then, why any job at all would be appealing, even if the daily wage is less than what it costs to buy a latte in other parts of the world. Working in a cobalt mine should provide some security, at least, and be reasonably safe.

If only that were the case.

Recently, *The New York Times*[16] reported on workforce conditions in Congolese cobalt mines, focusing on the holdings China Molybdenum acquired from the American company Freeport-McMoRan and its Canadian partner in 2016. One of those mines is the giant Tenke Fungurume.

With the purchase of Tenke Fungurume, China gained access to some of the largest cobalt reserves in the world; the mine produces more cobalt than any other mine and, indeed, twice as much cobalt as any other *country*. The mine costs China Moly, as it's often called, USD2.65 billion, with more than half that amount coming from loans made by Chinese state-owned banks.

That deal and others made during what can only be called a buying spree followed a diplomatic courtship in the early 2000s between former DRC President Joseph Kabila and then-Chinese President Hu Jintao. In exchange for China promising to build infrastructure, schools, hospitals, universities, and 2,000 miles of roads, Kabila agreed in 2008 to provide Hu's government open admission to his nation's minerals—10 million tons of copper and more than 600,000 tons of cobalt. The deal was valued at USD6 billion. China would replace the U.S. and others who had long supported the DRC but were scared off by its tarnished human rights record and history of corruption, providing a much-needed economic boost to the DRC.

It wasn't long before Chinese-backed road modernization projects were underway. The Sino-Congolese partnership also constructed stadiums, built water treatment facilities, and installed solar street lamps across the region.[17]

Against this backdrop, it's easy to imagine how appealing it might be to work at Tenke Fungurume under its new Chinese ownership.

The fact is, during the first years, things at the mine weren't all that bad for the nearly 10,000 mineral processors, drivers, mechanics, welders, security guards, and general workers (though China Moly did replace some African workers with its own people, setting off an unexpected unemployment blip).

Little by little, though, things like safety began to erode. This was a huge departure from the Freeport-McMoRan "zero tolerance" days when safety was a priority and there had been just one death recorded in more than eight years. Under China Moly's ownership, safety took second place to cost-cutting measures. "Workers climbed into acid tanks to conduct repairs without checking the air quality," *The New York Times* reported. "Others drove bulldozers and other heavy equipment without training or did dangerous welding jobs without proper oversight." Employees were also told to keep quiet about their injuries, lest they ruin the company's reporting records.

Worse yet, *The Times* said, when safety inspectors found violations, they were offered bribes to look the other way. When they continued to press the issue, violence sometimes followed: "One safety officer said he was thrown to the ground by a worker he had called out for improperly using welding equipment. The man twisted his arm and broke his cellphone and work-issue camera."

The Guardian uncovered similar circumstances when they looked into conditions at DRC mines, including Tenke Fungurume. The undertaking was part of what they called an "attempt to track the cobalt supply chain from industrial mines through a number of refiners and battery makers to some of the world's leading electric car manufacturers, including Tesla, VW, Volvo, Renault and Mercedes-Benz."[18] *The Guardian* reporters interviewed Congolese workers who said they had been discriminated against, earning less than their Chinese counterparts. They were insulted, beaten, and their experience ignored by their supervisors.

"We're being treated in a very bad way by the Chinese. I'm a victim of assault myself. I was slapped across the face four times," said one worker. Another described feeling humiliated and embarrassed after sitting through a two-hour meeting in Chinese that was followed by a two-minute translation. "The way they are treating our people, you can't believe," he said. "We are just expecting them to have respect for human life, instead of using people like slaves."[19]

Clearly, this description does not indicate the kind of culture that makes for happy or loyal employees.

So why do the Africans on the Tenke Fungurume payroll stay there? It's certainly not for the salary: One worker told *The Guardian* he made about USD3.50 a day, although with overtime the total can reach the equivalent of USD4.82. And it's not the free lunch, either: The same worker reported that he gets just two small bread rolls and a carton of juice. There's no such thing as sick leave, and the idea of a vacation is laughable. Conditions are so exploitative, in fact, that in November 2021, human rights watchdog Rights and Accountability in Development (Raid) and Congolese lawyers from the Centre d'Aide Juridico-Judiciaire accused the cobalt industry of keeping workers in poverty.

And while there's enough finger-pointing to go around, it should be noted that some of the employees who told their story to *The New York Times* and *The Guardian* aren't directly employed by the mine. Instead, they work for one of its many subcontractors, which account for fully 70% of Tenke Fungurume's workers. As *The Guardian* reported, "the use of subcontractors can leave workers in an extremely precarious position: often hired on short-term contracts, or no contract at all, with limited benefits, low pay and the threat of termination always hanging over them."[20]

Thankfully, things aren't this bleak everywhere. At some of the country's other large-scale industrial mines, the pay is decent and safety regulations are enforced. But even Swiss Glencore depends on some subcontractors for what it calls specialist work, creating an opening for the kind of abuse seen at Tenke Fungurume, Congo Dongfang International Mining (which is owned by Huayou Cobalt, Zhejiang, China, and others.

What it comes down to is this: In the DRC, there just aren't many alternatives for workers who want a regular job with a wage that may be low but is at least somewhat dependable.

And it could be worse: Workers could be "artisanal" miners.

Cruel Realities in the Artisanal Mines

In the West, the word *artisanal* has a positive connotation; it's associated with small, hand-made batches of cheese, for example, or locally made breads or wine.

In the DRC, however, artisanal cobalt miners—so-called because they work informally, finding small quantities of ore, then sell directly to traders or intermediaries, most of them Chinese—couldn't be farther from that cozy, upscale image.

There are an estimated 110,000 to 150,000 artisanal miners working in the DRC. Of that number, about 40,000 are children—many as young as age 7. They use only the most basic tools, even their bare hands, to pick cobalt from the rocks and tailings discarded as by-products at mining concessions. The adult miners risk their lives digging primitive underground tunnels to reach ore on private mine property, and some have died when the tunnels collapsed or flooded. The DRC estimates that 20% of the cobalt currently being exported from the country comes from artisanal miners.[21]

Obviously, the mines don't sanction this. For the most part, the men, boys, and girls engaged in artisanal mining are considered trespassers, and China Moly, for one, doesn't stop by just giving them a stern warning to leave the premises. Feeling overrun, the company asked the government to start patrolling the area and hired private security; according to *The New York Times*, one artisanal miner near Tenke Fungurume died when a soldier shot him.[22] Government and security agency officials aren't above turning a blind eye to the situation, however, and have even demanded payment from artisanal miners for "unauthorized" access to the sites.

As for the child workers, Amnesty International said it's not unusual for them to work as many as 12 hours a day for a dollar or two. With no way of verifying the weight of the sacks of ore they collect or the grade of the cobalt they pick, they have to accept what the traders give them. It's not hard to imagine they're being treated unfairly.

By the way, for some boys and girls, those long workdays are on top of a full day of school, or the weekends or holidays. The DRC has free and compulsory education, but participating in artisanal mining is the only way many youngsters from impoverished families can afford the uniforms and supplies their schools require.

Despite these hardships and mistreatment, what keeps the artisan miners on the hunt for cobalt? Where the cobalt ore came from and whether it was obtained legally or not doesn't matter much to the traders, meaning there's a steady market. Traders buy the ore, then sell it to larger companies in the DRC for processing and export to China. Amnesty International said Congo Dongfang Mining International is one of the largest companies at the center of this trade.[23]

At this point, we have to ask: Do the downstream companies that source smelted cobalt from Congo Dongfang Mining International parent Huayou Cobalt know about the child labor violations in the supply chain? When Amnesty International contacted many of them—big consumer companies with well-known names, usually with zero-tolerance policies regarding child labor—it was clear they had failed in their due diligence. Others simply denied using Huayou Cobalt as a supplier, even though records indicate a relationship.[24]

The Opportunities Are There… Africa Must Seize Them

In addition to working for large industrial mines—some offering far better conditions and pay than others—or being an artisanal miner, there are other ways Congolese are employed in the cobalt trade. Some work as laborers, earning a fee from the landowner. Others have profit-sharing arrangements with mine owners or business arrangements with investors.

Still, what all this means is that average Africans remain at the bottom of the economic food chain harvested from their natural resources, consigned to the menial task of extracting cobalt when that's not where the money or opportunity is.

Some African companies are trying to change this, to shift from an extractive economy to a mining economy that addresses the entire lithium-ion battery value chain. Much of the activity is currently happening outside the DRC, though, with varying signs of success.

Early in 2020, for example, South Africa's Megamillion Energy Company announced it would be "Africa's first large-scale producer of lithium-ion batteries, making lithium-ion batteries in Africa for Africa's benefit first." No one can argue with a goal like that, which they said they'd achieve by building a USD2 billion plant that would employ 3,500 South Africans and produce 32GWh of cells per year by 2028.[25] Of course, at the time of their announcement, no one had envisioned a global pandemic that was just weeks away. Since the original press coverage, Megamillion has been relatively silent.

The news out of Zambia is slightly more optimistic. There, Chambishi Metals produces cobalt metals; in fact, it's one of the largest refiners of its kind in the world. It doesn't produce exportable batteries, however, and the operation has been plagued by supply interruptions. In 2020, owner Eurasian Resources Group Africa—which also owns the Boss, Frontier, and Comide cobalt mines in the DRC—took Chambishi offline temporarily when it couldn't procure enough raw material to keep production running.[26] Since 2018, under an agreement with Tanzania, Magnis Energy Technologies has been moving its Nachu graphite processing plant toward completion[27]; in 2019, Mozambique signed a joint venture with Australia's Battery Minerals and U.S. graphite processor Urbix to build a graphite purification plant in the Cabo Delgado province.[28]

Even the DRC seems determined to not be left behind. The nation is formalizing its bid to develop an industrial policy around battery manufacturing. Speaking in the capital city of Kinshasa in November 2021, Prime Minister Jean-Michel Sama Lukonde Kyenge announced the creation of a battery council that would pilot "the government's policy to develop a regional value chain around the electric battery industry." Lukonde announced the DRC would launch a special purpose investment vehicle to raise funds for the effort.[29]

The time seems more than right for the DRC. With more electric vehicles being produced worldwide, the IEA expects a cobalt shortage by 2030.[30] Accelerating its activities across the value chain should enable the DRC to help fill the gap.

Unsurprisingly, it faces competition, and not just from China: Europe and the U.S. are getting into the production race as well. A few years ago, Stockholm-based battery developer Northvolt raised USD1 billion for a battery gigafactory in Sweden;[31] the company also has an assembly plant in Poland. In March 2022, Volkswagen and Northvolt, with the backing of Goldman Sachs, announced plans for a similar facility in Germany. Tesla is also building its own gigafactory there. And just a few days before the Volkswagen announcement, Ford said it would build a gigafactory near Ankara, Turkey.[32]

As for the U.S., start-up EV battery maker, Sila Nanotechnologies Inc., recently raised USD590 million in funding, and manufacturer Romeo Power went public in 2020.[33] With the administration of U.S. President Joe Biden promising increased support for green energy initiatives, efforts from U.S. companies are likely to accelerate.

Avoiding the Resource Curse, v.2

Is it realistic to expect that Africa can enter the increasingly crowded EV value chain fray and come out on top? One hope is that we will be able to leverage our relationship with the EU, our largest trade partner, to gain ground. After all, while the European Green Deal's proposals to cut greenhouse gas emissions by 55% from 1990 levels by 2030 most certainly will affect African oil production, it creates a pathway for more—and more lucrative—use of our mineral resources.

But that's only if we can avoid another resource curse.

In my last book, I argued that Africa can put an end to its oil and gas resource curse and provided strategies for getting there. Here, the first step is for our minerals-producing nations to enact an enabling policy structure, one that will improve competitiveness, broaden the scope of regulated activities, create coalitions of stakeholders, and reform the way foreign operators do business. With the right framework and incentives, we can attract investment to move beyond unearthing commodities to a manufacturing economy.

But that's just part of the mission. We must create local content policies that ensure equity for Africans without driving away foreign investors. These policies should allow Africans to compete for diverse, well-paying jobs, from entry-level work to positions of leadership, thereby creating backward linkages and setting guidelines for working with African suppliers, service companies, and other local businesses. These policies should spell out the kinds of relationships foreign entities should be making with host communities, including making sure there are capacity-building and knowledge-sharing programs. Finally, these policies can protect our people from mistreatment.

We must also give women an equal chance to benefit—rather than denying them a potentially lucrative and fulfilling career, as has happened so often in the oil and gas sector. As a recent report from sustainability organization BSR noted, there are significant barriers to women in the male-dominated mining industry. Not only is the field culturally unwelcoming, but women are also constrained by limited education, lack of skills, and little access to information. In some cases, they face outright discrimination and harassment or, worse yet, even gender-based violence. Women also have to navigate work environments not designed with them in mind, meaning they lack even basic sanitation facilities. They struggle, the report said, "to avail themselves of economic opportunities in mining value chains."[34]

When we've put in place the right policies, overcome these challenges, and taken all these measures, only then will we have adequately positioned individuals, communities, businesses, and nations to fully reap the economic opportunities that our mineral resources can provide.

And rather than reinventing the wheel, we should be looking for homegrown positive examples of mineral mining policies that other African countries can build upon. South Africa, for example, has a highly detailed local content framework for its mining industry that, while not perfect, includes employment quotas at all levels. It also promotes capacity building by including mandatory expenditures on training, supplier development, and community development programs.[35]

It's Not Too Late

The World Economic Forum says a sustainable future hinges in part on battery storage. Batteries are the near-term driver to decarbonizing road transportation. There's no reason why the value-additive aspects of mining—including smelting, refining, cell assembly, and EV production—cannot take place in Africa, creating well-paying jobs at every level, entrepreneurial opportunities, and security for millions.

We can bring children off their knees, lift men out of hazardous, hand-dug mines, give women access to dignified employment, and create economic growth for our people.

Chapter 21

COULD NAMIBIA SERVE AS A TEMPLATE FOR ENERGY TRANSITION SUCCESS IN AFRICA?

This book is going to press in 2022, a year that started with a bang for Namibia.

For a long time, this southern African state wasn't a big player in the oil and gas scene. It was known to possess a large offshore natural gas field—Kudu, estimated to hold 37 billion cubic meters (1.3 trillion cubic feet) of gas in place (GIP).[1] But Kudu was a stranded asset, considered to be so difficult to develop profitably that it had remained untapped since its discovery in 1974.

Yes, 1974. Nearly 50 years ago!

Several development options have been discussed over the years, but none of these plans made any progress. Kudu remained so isolated that none of the international oil companies (IOCs) that invested in the project was able to find an affordable way to fulfill the Namibian government's dream of sending gas from the field to shore to serve as fuel for an 800-MW thermal power plant (TPP).

That's probably going to change soon, though.

We've been hearing one exciting announcement after another from Namibia during the last couple of years. One by one, companies like Vancouver-based Reconnaissance Energy Africa (ReconAfrica), Shell Oil, and TotalEnergies have made major discoveries.

As I've been writing this book, the working petroleum system that ReconAfrica discovered in the Kavango Basin—spanning northeast Namibia and northwest Botswana—in 2021 has been described as potentially "the largest play of the decade." This year, ReconAfrica is moving forward with plans to drill at least three new test wells to prove the system is commercially viable.

Meanwhile, in early January 2022, Shell struck oil at Graff-1, a deepwater exploration well drilled at the PEL 39 block in the Orange Basin. It did not reveal the size of its discovery immediately, but by late January, industry experts had started offering estimates of 250-300 million barrels.[2] In mid-February, Wood Mackenzie went even higher, saying in a briefing note to its clients that Graff-1 might contain between 500 million and 1 billion barrels of light crude, including 700 million barrels in recoverable reserves.[3] This is large enough to classify the site as a giant field, the consultancy said.

Shortly thereafter, in the second half of February, TotalEnergies came forth with news of a bigger discovery. It said it had found oil in Venus-1x, a wildcat exploration well drilled within the PEL 56 block in the Orange Basin. Once again, TotalEnergies didn't say much at first about the size of its new field, but the first estimate that surfaced was 400 million barrels, larger than the initial guess for Graff-1.

But the next round of figures practically soared into the stratosphere. Wood Mackenzie wrote in late February, in a confidential report cited by Upstream, that Venus-1x might hold more than 3.0 billion barrels of crude in recoverable reserves, up from the initial estimate of 1.5-2.0 billion barrels.[4] If this is true, the site will become the biggest oil discovery ever made in Sub-Saharan Africa.

This is great news for Namibia, for multiple reasons:

- The fields contain light crude oil, which commands a higher price than other grades because it is easier to refine and contains lower levels of sulfur.
- The volumes of oil are so large that the fields will be economical to develop.
- The economics of the projects look even more favorable because the discoveries were made at a time when oil and gas prices were high and rising.[5] Shell intends to take advantage of that fact by fast-tracking development and bringing the Graff-1 field onstream as quickly as possible,[6] perhaps as early as 2026. Meanwhile, Namibian Energy Minister Tom Alweendo said that he wanted TotalEnergies to work at a similarly rapid pace at Venus-1x.[7]
- Fast-tracking is now easier and cheaper than ever because of advances in marine technologies, such as SBM Offshore's development of the Fast4Ward® modular technique for the construction of floating production, storage, and offloading (FPSO) vessels to support exploration and development drilling.[8]
- The fields also contain significant amounts of associated gas.[9] The volumes in question may be large enough to support an export-oriented LNG project. They could also be sent to shore by pipeline—and by a stroke of geographic good fortune, it would be easy to build such a pipeline along a route that also offered access to the Kudu field.
- As exploration and development work moves forward, economies of scale and infrastructure networks will emerge—and they will have the potential to become even more efficient if TotalEnergies and Shell make additional discoveries. They could also expand to include other IOCs working in Namibia's offshore zone.

In short, Namibia is now in the enviable position of having large amounts of hydrocarbons—11 billion barrels of oil and 62.3 billion cubic meters (2.2 trillion cubic feet) of gas[10]—plus a good chance of being able to attract large amounts of investment, capital, technology, and expertise from IOCs that want to explore and develop that oil and gas. It should also be able to use the economies of scale and infrastructure networks that IOCs establish to bring their production ashore—that is, to deliver it to Namibia's domestic market.

In other words, now that the IOCs have good reasons to spend money in the offshore zone, they may be able to install the infrastructure needed to bring Kudu gas to Namibia after nearly 50 years of waiting.

This is exciting news! But it doesn't end there.

Big Hydrogen

Namibia doesn't just have potential as a source of conventional energy. It also has the potential to become a powerhouse in the realm of renewable energy.

As President Hage Geingob noted in an October 2021 article for the World Economic Forum (WEF), Namibia has a particular advantage with respect to solar power: "We have the potential to capture around 10 hours of strong sunlight per day for 300 days per year. As a result, Namibia has some of the highest solar irradiance potential of any country in Africa, which is sufficient to provide power for our people and our neighbors."[11]

According to Geingob, Namibia is ready to put this potential to work—not just for itself but also for other countries in the region. (Just imagine what it might mean for Eskom, South Africa's beleaguered national electricity company, to have another source of supply to draw on. No more load-shedding! No more rolling blackouts!)

What's more, the president says, Namibia also wants its solar potential to serve long-term decarbonization goals. He noted in the WEF article that the country could use solar energy for the purpose of producing green hydrogen fuel, with solar farms powering desalination plants that turn out the vast quantities of fresh water that can then serve as raw material for hydrolysis plants.

This potential has not gone unnoticed. Geingob wrote that as of October 2021, foreign investors had already presented proposals to the Namibian government for projects that would contribute 6,400 MW to the country's total generating capacity within the next decade. This is about 10 times its current consumption of power, which stands at approximately 640 MW. (And please notice that he said consumption, not production; this is important because Namibia is currently a net importer of electricity.)[12]

Namibia has also attracted attention from investors specifically interested in hydrogen. In November 2021, it named Germany's Hyphen Hydrogen Energy, a joint venture between Enertrag and Nicholas Holdings, as the winner of a competitive international tender for the right to build a large-scale green hydrogen complex in the Tsau-Khaeb national park.[13] Hyphen has now negotiated a contract that envisions the establishment of a vertically integrated facility near the coast. This facility will utilize wind- and solar-powered generating units to power a desalination plant that handles water from the ocean, as well as the hydrolysis plant that splits the water to produce hydrogen to be used as fuel. It may also include a green ammonia unit that will convert the hydrogen into fuel.

This isn't a pilot project with a pilot project's budget. The total value of the contract has been estimated at USD9.4 billion,[14] with USD4.4 billion going to the first stage, estimated to start up in 2026. Nor is it a pilot project in terms of scale. It's going to be huge—at least two orders of magnitude larger than the biggest hydrogen plants in operation as of the first half of 2022.[15] The hydrogen complex's wind parks and solar farms are going to have a power generating capacity of 2,000 MW in the first stage of operations, and that number will climb, reaching 5,000 MW of electricity generating capacity and 3,000 MW of electrolyzer capacity in the late 2020s. Hydrogen production will be equally large in scale, with the electrolysis plant's production peaking at 300,000 tons a year.

And this is just one project. Hyphen's assigned site in Tsau//Khaeb covers about 5,700 square kilometers (2,200 square miles), equivalent to 0.69% of Namibia's total surface area of 824,292 square kilometers (318,261 square miles). In other words, it is surely not the only place in the country that is suitable for the construction of hydrogen plants and the renewable generating facilities needed to sustain them. Others can be built.

Geingob and the members of his administration surely know this. That's why it's no surprise that Namibia joined five other African states in May 2022 to set up the Africa Green Hydrogen Alliance.[16] This alliance will work, with the support of the UN, towards the goal of kickstarting the formation of a global supply chain for green hydrogen.

Go Blue to Go Green

Here's where things get tricky.

We've come to know, in a relatively short amount of time, that Namibia's energy resources are both vast and diverse. The country has large amounts of oil and gas, as well as a large amount of renewable generating capacity. So what are we—in the collective sense—going to do about that?

There are plenty of environmental activists who already have an answer: Leave the oil and gas where it is and forsake all hydrocarbons in favor of renewable energy. That's the position of the Economic and Social Justice Trust (ESJT), a Namibian non-governmental organization (NGO) that said in February 2022 that it expected Shell's oil discovery at Graff-1 to have dire consequences.[17]

Rob Parker, a representative of the NGO, told a South African newspaper that offshore oil development had the potential to cause catastrophic damage to Namibia's tourism and fishing industries, while also doing little to improve domestic fuel supplies or create new jobs. He also said he was suspicious that Shell would find ways to make the local population shoulder the environmental and social costs of its exploration and development work.

A better alternative, Parker said, would be for Namibia to concentrate on renewable energy and decarbonization initiatives.

"At a time when there are financial incentives to go green, this nation blessed with abundant sunshine should be leveraging [its] ... competitive advantage, but instead we are going down a path that everyone else is leaving ... Our advantage is the sun and we have a small population in the country. The country is stable and peaceful, and we have an ideal environment for solar. If we could do solar, we will get funding from COP and other green organizations," he said.

It would be nice if ESJT's position were an outlier, but it's not. It is typical of the stances taken by many environmental groups.

And that's unfortunate. Because if I understand Parker and his like-minded cohorts correctly, the answer is for Namibia to ignore part of its own vast resource endowment, to forego the opportunity to build up its own economy, and to go begging to the UN in the hope that the great and powerful arbiters of virtue will be so impressed by one small country's moral fortitude that they will cover the bill for its solar panels. (And need I even ask how many of these public servants will be sitting in air-conditioned offices where the climate-control systems depend on electricity churned out by gas-fired TPPs?)

Do they really mean this? Truly? Because it's a course of action I can't imagine following when the following alternative exists:

- Start by **expanding offshore drilling** for oil and gas.

- **Produce as much oil and gas as possible**—as quickly as possible, to meet the current needs of the market—using modular and marine solutions to reduce costs as appropriate.
- Don't just keep the machinery running. Along the way, **optimize Namibia's legal regime** and **standardize contracts** to promote technology transfer, job training, local entrepreneurship, and partnership with local companies. Make a particular effort to **encourage women to seek jobs in the energy sector** and **promote women-owned businesses**. Establish **pragmatic local content standards** that balance the needs of oil and gas companies and Namibians. **Keep Namibia's complicated history in mind** to ensure that the new laws and standards promote equality and improve the economic lot of those who suffered most from the apartheid regime that ended in 1990.
- **Export oil and gas,** use the proceeds from sales to **set up a sovereign wealth fund** and make other domestic investments, and figure out the optimal way to split gas production between the export market and the domestic market. Also, when planning gas export projects—and by that, of course, I mean LNG projects—consider using a mix of onshore gas liquefaction plants and offshore floating LNG (FLNG) facilities to **maximize energy security.**
- **Pay attention to the onshore upstream sector as well as offshore mega-projects.** For example, look for ways to optimize the development of onshore oil and gas resources, such as the contract area assigned to ReconAfrica in the Kavango Basin. Now that the company knows it has discovered a working petroleum system there, it is hoping to progress to commercial development. If it is able to do so, it will face obstacles. Its reserves are not as well-positioned for export, as they are hundreds of kilometers inland. However, they could also support

regional initiatives that increase and diversify fuel supplies in southern Africa. If their resources are large enough, they could, for instance, serve as another source of gas for a new TPP that Zambia intends to build in its capital city, Lusaka. (And this gas would have the advantage of being much closer to the Zambian border than the gas from offshore fields!

- **Meanwhile, allocate as much gas as possible to the domestic market** for the following purposes:
  - Producing electricity, which will be valuable for multiple reasons, including alleviating high levels of energy poverty within Namibia and generating revenue from exports to South Africa and other neighboring countries.[18]
  - Producing LPG for use as a home heating and cooking fuel that is much cleaner and safer than traditional biomass-based fuels such as wood, charcoal, or dung, all of which contribute to the pollution that causes thousands of Africans to die every day.[19]
  - Setting up gas-based industries such as fertilizer and petro-chemical production.
  - Supporting the operation of large industrial concerns that can utilize gas as a fuel source.
  - Establishing a blue hydrogen economy.

I view the last piece as crucial: blue hydrogen within the context of domestic gasification.

Let me explain what I mean.

A Template for Success

In blue hydrogen projects, gas produces the electricity that runs the desalination plants that turn out clear water and the hydrolyzers that split the clear water into oxygen and hydrogen. Gas makes things easy because it's a known quantity. That is, we already know quite well what to do to maximize electricity production from gas, and that can be a good thing—because it frees up time, money, and other resources for work on new things, such as establishing the supply chain for hydrogen (and perhaps ammonia as well). And producers will need that time to find customers, make the necessary financing and credit arrangements, set up transportation, develop contracts, obtain insurance, ensure compliance with the relevant regulations and industry standards, establish relationships with service providers, and so on.

But many of the links in the supply chains they establish are likely to be portable. For example, the electrolysis plant in a hydrogen project does not have to rely on water from a desalination plant using electricity from a particular source. It can use water from anywhere.

Gas-fired power, solar, wind—it doesn't matter. From a purely utilitarian standpoint, a hydrogen project just needs water. This means that, within the boundaries of its contractual obligations, it can start with gas-fired power and phase the gas out in favor of wind and/or solar over time.

Alternatively, let's say that a Namibian hydrogen producer arranges to move production to market via specialized tanker trucks. Is it going to matter to the transportation company whether the hydrogen being loaded into the tanks is blue or green? No, it won't. The only thing that matters is filling the tanks and telling the drivers where to go. There's nothing remarkable about the possibility that, over the course of a few years, the drivers may shift from only picking up loads at the blue plant to only picking up loads at the green plant.

Meanwhile, there's another important part of the supply chain that can make a similar shift: pipelines.[20] Yes, pipelines.

Hydrogen can be transported by pipeline in much the same way that gas is.[21] What's more, gas pipelines can be part of a transition to hydrogen, in that they can first be used to carry a blend of natural gas and hydrogen in advance of a switch to pure hydrogen. (Indeed, doing so would help establish a market for hydrogen and give pipeline operators an incentive to start addressing the technical challenges involved in transporting hydrogen and, to a lesser extent, hydrogen/gas mixtures.[22])

What all this means, in short, is that Namibia can prepare for a green hydrogen economy by building up its domestic gas sector. Gas development justifies the construction of the pipelines that will carry future hydrogen output while also powering the electrolysis plants used for blue hydrogen manufacturing. Gas development also fosters the creation of a supply chain with links that are ready to serve either blue or green hydrogen producers.

Gas can lay the groundwork for the future while also paying for itself in the present.

I think this is a better alternative for Namibia than leaving the oil and gas in the ground and hoping that proper concern for the climate will light the way as well as electricity does. I'd rather see the country use everything it has—not just the oil, not just the gas, not just the renewables, but all of them, and all of them at the right time—by treating them as complementary components of the same tool kit and not as antagonists.

If Namibia does that, it will create a brand-new template for energy transition success that the rest of Africa can follow. And I don't just mean financial success, I mean success on a human level. In this type of success, people have opportunities to lift themselves out of poverty, obtain an education, overcome obstacles, and find creative ways to make the most of their own talents.

Namibia has an opportunity to show the world what Africa can do. I'd really like to see it happen.

CONCLUDING THOUGHTS

In May 2021, I learned that Africa Oil Week, the continent's largest oil and gas conference, would be moving from its traditional location in Cape Town, South Africa, to Dubai in the United Arab Emirates (UAE). Conference organizer, London-based Hyve Group Plc, said the move was temporary, a necessary shift because of COVID-19 restrictions in South Africa.

So, instead of picking *any other city in the African continent*, the conference organizers felt the best place to discuss, plan for, and make deals in African oil and gas was *somewhere other than Africa*.

This kind of decision-making—the suggestion that Africa is somehow unqualified to host an oil conference it had hosted for nearly 30 years, where countless energy deals were made—epitomizes the same mindset that African oil- and gas-producing states encounter when it comes to their energy transition. It's the same insulting, patronizing mindset that has led environmental groups and Western countries to suggest they know what's best for Africa as they pressure our leaders into energy transition decisions that aren't in Africa's best interests. It's the same mindset that, in part, inspired me to write this book about the critical importance of a just, Africa-first energy transition.

The decision to move Africa Oil Week out of Africa only strengthened my resolve to get my book's message out. I was more determined than ever to show why Africans must have the final say on their countries' natural resources. I was also determined to work with African oil and gas industry stakeholders to deliver an Africa-first conference, an event that focused on African opportunities and perspectives and took place *in Africa*. I knew I was going to be ridiculed and laughed at by many, especially the Hyve group team. There were racial epithets, and coded words used against us. It only fueled our determination to push on.

Months later, in November 2021, the African Energy Chamber successfully launched African Energy Week in Cape Town.

I'd like to explain what happened from our perspective—and the significance of that highly successful inaugural conference—but first, you should know more about the Africa Oil Week that I remember. Maybe if you understand how much it means to Africans to have a conference devoted to their best interests, you'll see how dismayed and frustrated we were about losing it.

A Shining Legacy

I consider Dr. Duncan Clarke a visionary. An oil industry expert. An African role model.

Clarke, who was born and raised in Zimbabwe (then Rhodesia) has written numerous books about Africa and its oil industry. He's the founder of Global Pacific & Partners, an advisory firm for the upstream sector based in the Netherlands. But perhaps his greatest legacy to our continent was the creation of Africa Oil Week (originally known as the Africa Upstream Conference) in 1994.

The idea of this international conference was to develop and promote African countries that had oil and gas reserves but lacked the connections and financial resources to successfully participate in the global petroleum industry. Clarke made a point not only of inviting oil industry majors to the conference, but also African businesses and entrepreneurs who likely had few chances to be part of a large, international oil event.

His vision was a success. Through Africa Oil Week, he connected Africans, Americans, Asians, and Europeans. He connected representatives from the upstream, midstream, and downstream sectors. Those who attended Africa Oil Week had the chance to meet with African oil and gas ministers and industry executives from around the globe, leaders from national oil companies, and the CEOs of independent exploration and production companies. At Africa Oil Week, participants could make deals, speak with representatives of banks and investment companies, and grow their networks.

At the heart of the conference was the understanding that the event was all about Africa. You would walk in, and everything pointed to our continent, from color schemes to the signage. The conference's legendary elephant sign, for example, showcased presentations and displays related to the year's biggest discoveries and opportunities.

I will forever be grateful to Duncan Clarke for his incredible legacy. But when I think about him now, it's with a sense of sorrow.

Even in his wildest dreams, he never could have imagined what would happen when he decided to sell Global Pacific & Partners' majority interest in Africa Oil Week to London-based ITE Group (now named Hyve Group Plc) in 2015. He never could have imagined that the deep respect for and understanding of Africa that Africa Energy Week was founded upon would no longer guide event decisions.

And never was that shift in priorities more evident than when Hyve Group decided to move Africa Oil Week to Dubai.

Reasonable Outrage

COVID-19, frankly, was a weak excuse to move the conference out of Africa.

Yes, in 2021, the entire world was grappling with COVID. It wasn't simply a problem for South Africa or our continent. Even if the organizers felt that Cape Town wasn't an ideal host city—a conclusion that African Energy Week 2021 proved false—we felt strongly that Hyve Group could have chosen from any number of other African cities hosting large-scale events that year.

We could have taken this event to Dakar, Senegal, where other large events have been held. President Macky Sall has been very supportive of opening the country and driving a thriving energy industry there. The conference could have been taken to Kigali, Rwanda; to Cairo, Egypt; to Abuja or Lagos, Nigeria; or many other African locations.

The push to move a major African event outside of Africa is not an isolated incident. Before the 2021 Africa Cup of Nations in my home country, Cameroon, I heard Western sports commentators and analysts say this event, the most prestigious football (soccer) competition in Africa, should not be hosted in Africa. They argued for moving the tournament to Qatar, or somewhere in Europe, because of COVID in Africa.

The narrative that Africa is not good enough to host world-class events is offensive. And we at the African Energy Chamber felt we had an obligation to show how wrong it is.

A New Event with a Broader Focus: Africa Energy Week

So we took on the challenge of replacing Africa Oil Week with African Energy Week.

The shift from "oil" to "energy" in the name represented a realization on our part that we could no longer limit ourselves to discussions about the oil industry. We needed to look at oil and gas, yes, but as part of a broad energy mix that also included renewables.

With African Energy Week, we could have a one-stop-shop for productive conversations about Africa's energy future. And as we promoted the conference and invited participants, our vision resonated throughout Africa and around the globe.

How's this for proof: Out of all of Africa's oil-producing states, not one of their energy ministers or the heads of their national oil companies went to Dubai. They all came to Cape Town instead, and they stood with us to promote the continent's energy industry.

We drew representatives from international companies like Total Energies, Chevron, CNPC, CNOOC, ENI, Marathon, and Exxon Mobil, as well as service companies like Schlumberger, Technip Energies, ChampionX, Halliburton, and Baker Hughes. And we had participation from people like Günter Nooke, who has been the personal representative in Africa for Germany's outgoing Chancellor Angela Merkel. Merkel has always been good with us. I think she liked our honest approach to energy even when she might disagree on some issues.

And, we had strong, constructive dialogs, not only about our natural resources, but also about promoting free markets and limited government. We talked about renewable energy, too. One of my favorite moments was when Congo's Minister of Hydrocarbons, Bruno Jean-Richard Itoua, called upon the African Energy Chamber to lead a green energy initiative—something we are happy to do. What's brilliant about that request is that it calls for African renewable energy solutions driven by Africans. That is exactly what needs to happen.

We cannot have Africa's renewable energy future defined and driven by someone who visits Africa from time to time and doesn't understand the continent. We can't let something this important be directed by someone who doesn't prioritize the continent's silent majority: the 600 million people in Sub-Saharan Africa living in the dark and the 900 million with no access to clean cooking fuels. It's time for these voices to be heard. Their needs must guide African energy decisions. They're the reason our 2021 African Energy Week theme was "Making Energy Poverty History."

Next Steps

And, the continent's silent majority is why we will continue to deliver African energy conferences that put Africans' needs first. We are following up with African Energy Week 2022, which will be held October 18-21 in Cape Town.

Yes, we will remain unapologetically in favor of strong, thriving oil and gas industries for African nations—but we'll work toward reducing greenhouse gas emissions as well. As I've said, we simply don't believe abandoning our oil and gas reserves is remotely in Africa's best interests.

We have an obligation to be good stewards of the environment, but we also have an obligation to ensure that we don't have clean air in the dark.

And, we have an obligation to ensure that there is justice: We must provide our people with the reliable energy they need to achieve better futures. There are so many young people in Africa with ambitions and hopes for better lives, like those of their counterparts in Western and Asian countries, where access to reliable energy isn't an issue.

We must provide the change and solutions that Africa needs so that we are no longer seen as the dark continent.

And we, Africans, must be the ones to drive those changes and solutions. We cannot be the first to go out there begging, asking for handouts every time there is a crisis. Just as we demand respect and insist on being heard, we must believe in ourselves, too, rather than looking to others to meet our needs.

Of course, we have challenges, but we only harm ourselves if we use them to excuse poor decisions. People around the globe call us out on past ways of doing things, on corruption. We must accept responsibility for our shortcomings and turn things around.

I truly believe that we can create an energy industry that is a force for good; that we can overcome negative perceptions of who we are now by what we do next.

Organizing African Energy Week has been a step in that direction. It shows that Africans can organize world-class events that people from America, Europe, and Asia are proud to be part of.

Many people doubted we could do it, but we exceeded everyone's expectations. By the time African Energy Week concluded, participants made nearly USD24 billion worth of deals.

And we can be equally successful when it comes to our energy transition. We can create a thriving energy industry that offers Africans a better future. And we can drive a just transition to renewables based on African needs and priorities. We're already on our way.

ENDNOTES

Preface: Remembering a Friend, a Mentor, and One of the African Energy Industry's Greatest Allies

[1]Samuel Stebbins. "These 15 countries, as home to largest reserves, control the world's oil." May 22, 2019. USA Today. https://www.usatoday.com/story/money/2019/05/22/largest-oil-reserves-in-world-15-countries-that-control-the-worlds-oil/39497945

[2]Summer Said and Benoit Faucon. "OPEC Secretary-General Mohammed Barkindo Dies." The Wall Street Journal. July 6, 2022. https://www.wsj.com/articles/opec-secretary-general-mohammed-barkindo-dies-11657090828?page=1

[3]Summer Said and Benoit Faucon. "OPEC Secretary-General Mohammed Barkindo Dies." The Wall Street Journal. July 6, 2022. https://www.wsj.com/articles/opec-secretary-general-mohammed-barkindo-dies-11657090828?page=1

[4]HE Mohammad Sanusi Barkindo: Secretary General of OPEC. OPEC. https://www.opec.org/opec_web/static_files_project/media/downloads/about_us/HE%20Barkindo%20CV%20(10052019).pdf

[5]David Hobbs. "Remembering OPEC Secretary-General Mohammad Sanusi Barkindo." EnergySource-Atlantic Council. July 7, 2022. https://www.atlanticcouncil.org/blogs/energysource/remembering-opec-secretary-general-mohammad-sanusi-barkindo

[6]Amanda Battersby. "'Genuine Opec icon': Industry mourns the death of Mohammad Sanusi Barkindo." Upstream. July 8, 2022. https://www.upstreamonline

.com/people/-genuine-opec-icon-industry-mourns-the-death-of-mohammad-sanusi-barkindo/2-1-1253836

[7]Oluwatoyin Bayagbon. "Barkindo to receive 'Africa Oil Man of the Year' award." The Cable. May 15, 2018. https://www.thecable.ng/barkindo-receive-africa-oil-man-year-award

[8]"OPEC Secretary General Says Africa Needs to Combat Energy Poverty As a Priority." Zawya. August 24, 2021. https://www.zawya.com/en/press-release/opec-secretary-general-says-africa-needs-to-combat-energy-poverty-as-a-priority-r6mklu5k

[9]Sunnex Egbu. "Late Barkindo was a worthy ambassador of Nigeria – Sylva." NaijaOnPoint. July 7, 2022. https://naijaonpoint.com.ng/late-barkindo-was-a-worthy-ambassador-of-nigeria-sylva

[10]"Sylva Congratulates Barkindo." Swift Reporters. June 2022. https://swiftreporters.com/sylva-congratulates-barkindo

[11]NJ Ayuk. "Marking the End of An Era: OPEC Secretary General Barkindo Served with Distinction." The Guardian. January 13, 2022. https://guardian.ng/apo-press-releases/marking-the-end-of-an-era-opec-secretary-general-barkindo-served-with-distinction-by-nj-ayuk

Chapter 1: Africa Needs Energy—and More Than One Kind of Energy

[1]Frans Timmermans and Fatih Birol. "Time to Make Energy Poverty in Africa a Thing of the Past." European Commission. June 17, 2021. https://ec.europa.eu/commission/commissioners/2019-2024/timmermans/announcements/time-make-energy-poverty-africa-thing-past_en

[2]Samantha Gross. "Why Are Fossil Fuels So Hard to Quit?" Brookings. June 2020. https://www.brookings.edu/essay/why-are-fossil-fuels-so-hard-to-quit/

[3]"What Is the Paris Agreement?" United Nations Climate Change. n.d. https://unfccc.int/process-and-meetings/the-paris-agreement/the-paris-agreement

[4]"Responding to Climate Change." United Nations Environment Programme. n.d. https://www.unep.org/regions/africa/regional-initiatives/responding-climate-change

[5]Frans Timmermans and Fatih Birol. "Time to Make Energy Poverty in Africa a Thing of the Past." European Commission. June 17, 2021. https://ec.europa.eu/commission/commissioners/2019-2024/timmermans/announcements/time-make-energy-poverty-africa-thing-past_en

[6]Hannah Ritchie and Max Roser. "Access to Energy." Our World in Data. 2020. https://ourworldindata.org/energy-access#:~:text=940%20million%20(13%25%20of%20the,not%20have%20access%20to%20electricity

[7]"What African City Climate Action Looks Like." CDP. 2022. https://www.cdp.net/en/research/global-reports/africa-report

[8]Julie Bos, Lorena Gonzalez, and Joe Thwaites. "Are Countries Providing Enough to the $100 Billion Climate Finance Goal?" World Resources Institute. October 7, 2021. https://www.wri.org/insights/developed-countries-contributions-climate-finance-goal

[9]Jonathan Wilkinson and Jochen Flasbarth. "Climate FInance Delivery Plan: Meeting the US$100 Billion Goal." 2021. https://ukcop26.org/wp-content/uploads/2021/10/Climate-Finance-Delivery-Plan-1.pdf

[10]NJ Ayuk. "Greenpeace and Western Anti Africa Energy Groups Take the Wrong Approach to East Africa Crude Oil Pipeline (EACOP) in Uganda and Tanzania." African Energy Chamber. April 18, 2021. https://energychamber.org/greenpeace-and-western-anti-africa-energy-groups-take-the-wrong-approach-to-east-africa-crude-oil-pipeline-eacop-in-uganda-and-tanzania

[11]Nick Cameron and Mike Rego. "Underexplored African Source Rocks." GeoExPro. 2022. https://www.geoexpro.com/articles/2020/11/underexplored-african-source-rocks

[12]Matthew Goosen. "Top 10 African Countries Sitting on the Most Natural Gas." Energy Capital & Power. 2022. https://energycapitalpower.com/top-ten-african-countries-sitting-on-the-most-natural-gas

[13]"How Nigeria 'Accidentally' Discovered 206trn Cubic Feet Gas Reserve – Sylva." Premium Times. June 27, 2021. https://www.premiumtimesng.com/news/top-news/470180-how-nigeria-accidentally-discovered-206trn-cubic-feet-gas-reserve-sylva.html

[14]Matthew Goosen. "Top 10 African Countries Sitting on the Most Natural Gas." Energy Capital & Power. 2022. https://energycapitalpower.com/top-ten-african-countries-sitting-on-the-most-natural-gas

[15]Andries Rossouw, James Mackay, Christie Viljoen, Jeremy Dore, Matthew Plumstead, Pearl Maseko, Stacey Buck, and Wessel van Wyk. "African Energy Review 2021." PWC. November 2021. https://www.pwc.com/ng/en/assets/pdf/africa-energy-review-2021.pdf

[16]"Electric Power Sector CO2 Emissions Drop as Generation Mix Shifts From Coal to Natural Gas." US Energy Information Administration. June 9, 2021. https://www.eia.gov/todayinenergy/detail.php?id=48296

[17]Mark Green. "Eight Points – Natural Gas, Reducing Emissions and Environment." American Petroleum Institute. July 12, 2021. https://www.api.org/news-policy-and-issues/blog/2021/07/12/eight-points-natural-gas-reducing-emissions-and-environment

[18]Eric Kumeh Mensah. "Why Efforts to Clean up Charcoal Production in Sub-Saharan Africa Aren't Working." The Conversation. February 23, 2021. https://theconversation.com/why-efforts-to-clean-up-charcoal-production-in-sub-saharan-africa-arent-working-153462

[19]"Why Is the Switch to Gas for Cooking Still so Slow in Africa?" New African. December 10, 2020. https://newafricanmagazine.com/24602

[20]Robert Bryce. "The Real Problem With Renewables." Forbes. May 11, 2020. https://www.forbes.com/2010/05/11/renewables-energy-oil-economy-opinions-contributors-robert-bryce.html?sh=5cf105621403

[21]"What Is 'Intermittency' in Renewable Energy?" EnergyX. n.d. https://energyx.com/resources/what-is-intermittency-in-renewable-energy

[22]Aaron Ross. "Cobalt-Rich Congo Tries to Push into Battery Manufacturing" Financial Post. November 24, 2021. https://financialpost.com/pmn/business-pmn/cobalt-rich-congo-tries-to-push-into-battery-manufacturing

Chapter 2: More Harm Than Good: How a Rushed Energy Transition Can Hurt Africa

[1]Sam Meredith. “Europe’s Energy Crisis Is Making the Market Nervous. And Analysts Expect Record-High Prices to Persist.” CNBC. September 16, 2021. https://www.cnbc.com/2021/09/16/europes-energy-crisis-is-making-the-market-nervous-ahead-of-winter.html

[2]Chloe Taylor. “British Energy Firms Fear Collapse as Europe’s Gas Crisis Sees Prices Surge 250%.” September 20, 2021. CNBC. https://www.cnbc.com/2021/09/20/british-energy-firms-fear-collapse-as-european-gas-prices-surge-250percent.html

[3]Sam Meredith. “Europe’s Energy Crisis Is Making the Market Nervous. And Analysts Expect Record-High Prices to Persist.” CNBC. September 16, 2021. https://www.cnbc.com/2021/09/16/europes-energy-crisis-is-making-the-market-nervous-ahead-of-winter.html

[4]Baker Institute (Twitter post). September 21, 2021. https://twitter.com/BakerInstitute/status/1440323182125142032?ref_src=twsrc%5Etfw%7Ctwcamp%5Eembeddedtimeline%7Ctwterm%5Eprofile%3AEuropeanGasHub%7Ctwgr%5EeyJ0ZndfZXhwZXJpbWVudHNfY29va2llX2V4cGlyYXRpb24iOnsiYnVja2V0IjoxMjA5NjAwLCJ2ZXJzaW9uIjpudWxsfSwidGZ3X2hvcml6b25fdHdlZXRfZW1iZWRfOTU1NSI6eyJidWNrZXQiOiJodGUiLCJ2ZXJzaW9uIjpudWxsfSwidGZ3X3NwYWNlX2NhcmQiOnsiYnVja2V0Ijoib2ZmIiwidmVyc2lvbiI6bnVsbH19%7Ctwcon%5Etimelinechrome&ref_url=https%3A%2F%2Fwww.europeangashub.com%2Ftag%2Fdutch-ttf

[5]Chloe Taylor. “British Energy Firms Fear Collapse as Europe’s Gas Crisis Sees Prices Surge 250%.” September 20, 2021. CNBC. https://www.cnbc.com/2021/09/20/british-energy-firms-fear-collapse-as-european-gas-prices-surge-250percent.html

[6]Stanley Reed. “To Reach Climate Goals, New Oil and Gas Investment Must Be Stopped, Energy Agency Says.” The New York Times. May 18, 2021. https://nyti.ms/3ipevMV

[7]“How Much Carbon Dioxide Is Produced When Different Fuels Are Burned?” US Energy Administration. May 10, 2022. https://www.eia.gov/tools/faqs/faq.php?id=73&t=11

[8]Gyude Moore and Vijaya Ramachandran. "Economic Growth in Africa Will not Be Achieved by a Blanket Ban on Fossil Fuels." The Hill. March 4, 2021. https://thehill.com/blogs/congress-blog/foreign-policy/541554-economic-growth-in-africa-will-not-be-achieved-by-a

[9]Gyude Moore and Vijaya Ramachandran. "Economic Growth in Africa Will not Be Achieved by a Blanket Ban on Fossil Fuels." The Hill. March 4, 2021. https://thehill.com/blogs/congress-blog/foreign-policy/541554-economic-growth-in-africa-will-not-be-achieved-by-a

[10]"What Is ESG? Definition and Meaning." Market Business News. n.d. https://marketbusinessnews.com/financial-glossary/esg-definition-meaning

[11]"The State of African Energy 2022." African Energy Chamber. 2022. https://energychamber.org/wp-content/uploads/AEC-Outlook-2022_b.pdf

[12]David Sheppard. "Pandemic Crisis Offers Glimpse Into Oil Industry's Future." Financial Times. May 2, 2020. https://www.ft.com/content/99fc40be-83aa-11ea-b872-8db45d5f6714

[13]"Shell: Netherlands Court Orders Oil Giant to Cut Emissions." BBC. May 26, 2021. https://www.bbc.com/news/world-europe-57257982

[14]"Shell: Netherlands Court Orders Oil Giant to Cut Emissions." BBC. May 26, 2021. https://www.bbc.com/news/world-europe-57257982

[15]"Our Climate Target: Shell's Target Is to Become a Net-Zero Emissions Energy Business by 2050." Shell. n.d. https://www.shell.com/energy-and-innovation/the-energy-future/our-climate-target.html#iframe=L3dlYmFwcHMvY2xpbWF0ZV9hbWJpdGlvbi8

[16]Barry Po. "An ESG Reckoning Has Arrived For The Oil And Gas Industry." Forbes. August 3, 2021. https://www.forbes.com/sites/forbestechcouncil/2021/08/03/an-esg-reckoning-has-arrived-for-the-oil-and-gas-industry/?sh=597293fd51fb

[17]Elizabeth Corner. "Sub-Saharan Africa's Oil Production Goals Could Be Harmed by Lack of Investment and New Projects." World Pipelines. August 6, 2021. https://www.worldpipelines.com/project-news/06082021/sub-saharan-africas-oil-production-goals-could-be-harmed-by-lack-of-investment-and-new-projects/

[18]Eklavya Gupte. "African Energy Chamber Calls for Boycott of Firms Shunning Africa's Oil Sector." S&P Global. July 13, 2021. https://www.spglobal.com/platts/en/market-insights/latest-news/natural-gas/071321-african-energy-chamber-calls-for-boycott-of-firms-shunning-africas-oil-sector

[19]"The World Bank in Uganda." The World Bank. April 17, 2022. https://www.worldbank.org/en/country/uganda/overview#1

[20]"Total to Take Steps to Mitigate Environmental Impact of Uganda, Tanzania Projects." Reuters. March 8, 2021. https://www.reuters.com/article/total-uganda-environment-idCNL1N2L60K3

[21]"TILENGA & EACOP: Projects With a Socio-Economic Interest for Uganda and Tanzania." Total. n.d. https://totalenergies.com/sites/g/files/nytnzq121/files/atoms/files/uganda-projects_introduction.pdf

[22]"Uganda Oil Revenue Management: Closing Gaps in the Fiscal and Savings Frameworks to Maximize Benefits." The World Bank. 2020. https://openknowledge.worldbank.org/handle/10986/33899

[23]"Uganda: Government and Society." Brittanica. n.d. https://www.britannica.com/place/Uganda/Government-and-society

[24]"The World Bank in Uganda." The World Bank. April 17, 2022. https://www.worldbank.org/en/country/uganda/overview#1

[25]"The World Bank in Uganda." The World Bank. April 17, 2022. https://www.worldbank.org/en/country/uganda/overview#1

[26]"The Race for Oil and Gas in Africa." Al Jazeera. October 23, 2016. https://www.aljazeera.com/news/2016/10/23/the-race-for-oil-and-gas-in-africa

[27]Anton Botes, Andrew Lane, and Hannah Marais. "The New Frontier: Winning in the African Oil and Gas Industry." March 15, 2019. Deloitte. https://www2.deloitte.com/us/en/insights/industry/oil-and-gas/africa-oil-gas-industry-energy-reserves.html

[28]Elliot Smith. "Oil Nations Tipped for Political Instability if the World Moves Away From Fossil Fuels." CNBC. March 26, 2021. https://www.cnbc.com/2021/03/26/oil-nations-tipped-for-political-instability-if-fossil-fuels

-abandoned.html "Mozambique: Civilians Killed as War Crimes Committed by Armed Group, Government Forces, and Private Military Contractors – New Report." Amnesty International. March 2, 2021. https://www.amnesty.org/en/latest/news/2021/03/mozambique-civilians-killed-as-war-crimes-committed-by-armed-group-government-forces-and-private-military-contractors-new-report/

[29]"Mozambique: Civilians Killed as War Crimes Committed by Armed Group, Government Forces, and Private Military Contractors – New Report." Amnesty International. March 2, 2021. https://www.amnesty.org/en/latest/news/2021/03/mozambique-civilians-killed-as-war-crimes-committed-by-armed-group-government-forces-and-private-military-contractors-new-report/

[30]"Mozambique Insurgency: Children Beheaded, Aid Agency Reports." BBC. March 16, 2021. https://www.bbc.com/news/world-africa-56411157

[31]Ashoka Mukpo. "Gas Fields and Jihad: Mozambique's Cabo Delgado Becomes a Resource-Rich War Zone." Mongabay. 26 April 26, 2021. https://news.mongabay.com/2021/04/gas-fields-and-jihad-mozambiques-cabo-delgado-becomes-a-resource-rich-war-zone

[32]"Mozambique Gas Project: Total Halts Work After Palma Attacks." BBC. April 26, 2021. https://www.bbc.com/news/world-africa-56886085

[33]"The State of African Energy 2022." African Energy Chamber. 2022. https://energychamber.org/wp-content/uploads/AEC-Outlook-2022_b.pdf

[34]Vijaya Ramachandran. "Blanket Bans on Fossil-Fuel Funds Will Entrench Poverty." April 20, 2021. Nature. https://www.nature.com/articles/d41586-021-01020-z

Chapter 3: The Realities of Energy Poverty in Africa

[1]Khameer Kidia. "My Family in Zimbabwe Will Be Among the Last in the World to Receive the Vaccine." Slate. December 29, 2020. https://slate.com/technology/2020/12/covid-vaccine-priority-inequality-zimbabwe.html

[2]"Access to Modern Energy: Basic Energy Access on Household Level." energypedia. January 8, 2019. https://energypedia.info/wiki/Access_to_Modern_Energy#Basic_Energy_Access_on_Household_Level

[3]"Access to Modern Energy: Distribution of Access." energypedia. January 8, 2019. https://energypedia.info/wiki/Access_to_Modern_Energy#Distribution_of_Access

[4]Philip Gordon. "South Africa: 12-hour Rolling Blackouts on the Cards, Says Utility Eskom." Smart Energy International. January 14, 2020. https://www.smart-energy.com/industry-sectors/energy-grid-management/south-africa-12-hour-rolling-blackouts-on-the-cards-says-utility-eskom/

[5]Sri Mulyani Indrawati. "What You Need To Know About Energy and Poverty." World Bank Blogs. July 28, 2015. https://blogs.worldbank.org/voices/what-you-need-know-about-energy-and-poverty

[6]Genevieve McInnes. "Understanding the Distributional and Household Effects of the Low-Carbon Transition in G20 Countries." February 24, 2017. https://www.oecd.org/environment/cc/g20-climate/collapsecontents/McInnes-distributional-and-household-effects-low-carbon-transition.pdf

[7]Stefan Bouzarovski and Saska Petrova. "A Global Perspective on Domestic Energy Deprivation: Overcoming the Energy Poverty–Fuel Poverty Binary." Energy Research & Social Science. November 2015. https://www.sciencedirect.com/science/article/pii/S221462961500078X#bib1195

[8]"5 Facts You Should Know About Energy Poverty." FINCA. 2021. https://finca.org/blogs/5-facts-you-should-know-about-energy-poverty/?gclid=Cj0KCQiAqdP9BRDVARIsAGSZ8AleffZTA2QJ3eEwjJOBiOhT2mNcryG7-ITYJwY3XUZyrd9nOTTqpoAaAt67EALw_wcB

[9]"5 Facts You Should Know About Energy Poverty." FINCA. 2021. https://finca.org/blogs/5-facts-you-should-know-about-energy-poverty/?gclid=Cj0KCQiAqdP9BRDVARIsAGSZ8AleffZTA2QJ3eEwjJOBiOhT2mNcryG7-ITYJwY3XUZyrd9nOTTqpoAaAt67EALw_wcB

[10]"Access to Modern Energy: Clean Water." energypedia. January 8, 2019. https://energypedia.info/wiki/Access_to_Modern_Energy#Clean_Water

[11]"Sanitation." World Health Organization. March 21, 2022. https://www.who.int/news-room/fact-sheets/detail/sanitation

[12]Sri Mulyani Indrawati. “What You Need To Know About Energy and Poverty.” World Bank Blogs. July 28, 2015. https://blogs.worldbank.org/voices/what-you-need-know-about-energy-and-poverty

[13]“Access to Modern Energy: Livelihood Benefits.” energypedia. January 8, 2019. https://energypedia.info/wiki/Access_to_Modern_Energy#Livelihood_Benefits

[14]Leo Iacovone, Vijaya Ramachandran, and Martin Schmidt. “Stunted Growth: Why Don’t African Firms Create More Jobs?” Center for Global Development. February 3, 2014. https://papers.ssrn.com/sol3/papers.cfm?abstract_id=2390897

[15]Anne Nyambane. “Beyond Construction: The Challenges of Deploying Mini-Grids in Kenya.” Chatham House. June 18, 2019. https://www.chathamhouse.org/2019/06/beyond-construction-challenges-deploying-mini-grids-kenya

[16]“Africa Energy Outlook 2019.” International Energy Agency. November 2019. https://www.iea.org/reports/africa-energy-outlook-2019

Chapter 4: Foreign Aid to Africa: It’s Time to End This Harmful Cycle

[1]Anthony Leddin. “Foreign Aid Should be a Hand Up, Not a Handout.” Seed World. April 30, 2020. https://seedworld.com/foreign-aid-should-be-a-hand-up-not-a-handout/

[2]Jonathan Lea. “Why Foreign Aid Is Harmful.” Jonathan Lea Network. May 25th, 2022. https://www.jonathanlea.net/2015/why-foreign-aid-is-harmful

[3]Nathan Nunn and Nancy Qian. “The Determinants of Food Aid Provisions to Africa and the Developing World.” In Edwards, Johnson, Weil: African Successes: Sustainable Growth. Vol. IV. 2016. https://scholar.harvard.edu/nunn/publications/determinants-food-aid-provisions-africa-and-developing-world

[4]“Declining Aid, Rising Debt Thwarting World’s Ability to Fund Sustainable Development, Speakers Warn at General Assembly High-Level Dialogue.” United Nations Meetings Coverage and Press Releases. September 26, 2019. https://www.un.org/press/en/2019/ga12191.doc.htm

[5]Address by Nelson Mandela for the "Make Poverty History" Campaign, London - United Kingdom. February 3, 2005. http://www.mandela.gov.za/mandela_speeches/2005/050203_poverty.htm

[6]"Cultivating New Frontiers: 2021 Annual Report." One Acre Fund. 2022. https://oneacrefund.org/2021-annual-report/

[7]"Uganda." World Food Program USA. 2022. https://www.wfpusa.org/countries/uganda

[8]"Module 9: Food Aid Program Development." Unite for Sight. n.d. https://www.uniteforsight.org/effective-program-development/module9

[9]"Power Africa Initiative." African Development Bank Group. n.d. https://www.afdb.org/en/topics-and-sectors/initiatives-partnerships/power-africa-initiative

Chapter 5: Oil and Gas Discoveries We Can't Ignore

[1]"What is the Kavango Basin?" Petro Online. January 23, 2021. https://www.petro-online.com/news/fuel-for-thought/13/breaking-news/what-is-the-kavango-basin/54209

[2]David McKenzie and Ingrid Formanek. "A Canadian Oil Firm Thinks it Has Struck Big. Some Fear it Could Ravage a Climate Change Hotspot." CNN. May 3, 2021. https://www.cnn.com/2021/05/03/africa/namibia-oil-exploration-intl-cmd/index.html

[3]David McKenzie and Ingrid Formanek. "A Canadian Oil Firm Thinks it Has Struck Big. Some Fear it Could Ravage a Climate Change Hotspot." CNN. May 3, 2021. https://www.cnn.com/2021/05/03/africa/namibia-oil-exploration-intl-cmd/index.html

[4]"TotalEnergies Makes Large Oil Discovery off Namibia. Reuters. February 24, 2022. https://www.reuters.com/business/energy/totalenergies-makes-large-hydrocarbon-discovery-off-namibia-sources-2022-02-24

[5]Leo Holtz, Sakinatou Djantchiemo, and Abraham White. "Africa in the News: Major Oil and Gas Discovery in Namibia; Coup d'Etat in Burkina Faso; and COVID Update." Brookings. January 29, 2022. https://www.brookings.edu/blog

/africa-in-focus/2022/01/29/africa-in-the-news-major-oil-and-gas-discovery-in-namibia-coup-detat-in-burkina-faso-and-covid-update

[6]"Net Zero by 2050." International. May 2021. Energy Agency https://www.iea.org/reports/net-zero-by-2050

[7]Yuka Obayashi and Sonali Paul. "Asia Snubs IEA's Call To Stop New Fossil Fuel Investments." Reuters. May 19, 2021. https://www.reuters.com/business/energy/asia-snubs-ieas-call-stop-new-fossil-fuel-investments-2021-05-19/

[8]"Why Norway's government says ending oil industry 'won't help' cut global emissions," The Local, February 6, 2020. https://www.thelocal.com/20200206/why-norways-government-says-discontinuing-oil-production-wont-help-cut-global-emissions

[9]Yuka Obayashi and Sonali Paul. "Asia Snubs IEA's Call To Stop New Fossil Fuel Investments." Reuters. May 19, 2021. https://www.reuters.com/business/energy/asia-snubs-ieas-call-stop-new-fossil-fuel-investments-2021-05-19

[10]Matthew Goosen. "Angola Crude Oil Exports Increased by 3% During Q4 2021." Energy Capital & Power. n.d. https://energycapitalpower.com/angola-crude-oil-exports-increased-by-3-during-q4-2021

[11]"Ivory Coast: Italian Oil Giant Eni Makes Huge Offshore Discovery." DW. n.d. https://www.dw.com/en/ivory-coast-italian-oil-giant-eni-makes-huge-offshore-discovery/a-59056939

[12]Baudelaire Mieu. "Côte d'Ivoire: Eni's Mega Oil Discovery Revives the Ouattara Government's Ambitions." The Africa Report. September 3, 2021. https://www.theafricareport.com/124260/cote-divoire-enis-mega-oil-discovery-revives-the-ouattara-governments-ambitions

[13]Christophe Le Bec. "Côte d'Ivoire: Why ENI's Discovery Is a Game-Changer for West Africa." The Africa Report. November 29, 2021. https://www.theafricareport.com/151631/cote-divoire-why-enis-discovery-is-a-game-changer-for-west-africa

[14]Nermina Kulovic. "New Oil Discovery in Gulf of Suez Is One of Area's Largest in 20 Years." Offshore Energy. February 16, 2022. https://www.offshore-energy.biz/new-oil-discovery-in-gulf-of-suez-is-one-of-areas-largest-in-20-years

[15]"Senegal Enters the LNG Race With Significant Offshore Discoveries." Power Engineering. October 31, 2019. https://www.power-eng.com/gas/senegal-enters-the-lng-race-with-significant-offshore-discoveries

[16]Par Edialo. "Senegal Will Deploy a Floating Storage and Regasification Unit Near Its Coast in the Coming Weeks." Africa Logistics Magazine. March 18, 2021. https://www.africalogisticsmagazine.com/?q=en/content/lng-senegal-will-deploy-floating-storage-and-regasification-unit-near-its-coast-coming-weeks

[17]Bate Felix. "Total's South Africa Discovery Could Hold 1 Billion Barrels Oil Equivalent: CEO." Reuters. February 7, 2019. https://www.reuters.com/article/us-total-exploration-discovery/totals-south-africa-discovery-could-hold-1-billion-barrels-oil-equivalent-ceo-idUSKCN1PW0LF

[18]Charné Hollands. "Five Major Oil Discoveries in Africa in 2021." Capital Energy & Power. N.d. https://energycapitalpower.com/five-major-oil-discoveries-in-africa-in-2021

[19]"Eni Announces a Significant Oil Discovery in Block 4, Offshore Ghana." Eni. July 6, 2021. https://www.eni.com/en-IT/media/press-release/2021/07/eni-announces-significant-discovery-block-ghana.html

[20]"With Afina Unitisation, Ghana Gives Explorer Springfield E&P a Boost and Ensures Production Growth." APO Group. November 12, 2020. https://menafn.com/1101116940/With-Afina-Unitisation-Ghana-Gives-Explorer-Springfield-EP-a-Boost-and-Ensures-Production-Growth

[21]"Harlequin Oil and Gas Company To Sponsor More Development Programmes." Ghana Web. November 15, 2020. https://www.ghanaweb.com/GhanaHomePage/business/Harlequin-Oil-and-Gas-Company-to-sponsor-more-development-programmes-1109188

Chapter 6: The Vital Importance of Natural Gas

[1]"What's the Difference Between LPG and LNG?" Petro Online. October 7, 2019. https://www.petro-online.com/news/analytical-instrumentation/11/breaking-news/whats-the-difference-between-lpg-and-lng/50495

[2]"Assessing Potential for BioLPG Production and Use within the Cooking Energy Sector in Africa." The Global LPG Partnership. September 3, 2020. https://static1.squarespace.com/static/5633c4c2e4b05a5c7831fbb5/t/5f63e6ff27c2823bf0e2028b/1600382727754/GLPGP+-+Potential+for+BioLPG+Production+and+Use+as+Clean+Cooking+Energy+in+Africa

[3]"Xpressgas Limited Wins 2 Prestigious Awards at the Ghana Oil and Gas Awards 2019." Xpress Gas. December 14, 2019. https://xpressgas.com/xpressgas-limited-wins-two-2-prestigious-awards-at-the-ghana-oil-and-gas-awards-2019

[4]"Investing in Natural Gas for Africans: Doing Good and Doing Well." Africa50 / Energy Futures Initiative. 2018. https://www.africa50.com/fileadmin/uploads/africa50/Documents/Knowledge_Center/Investing_in_Natural_Gas_for_Africans_-_Doing_Good_and_Doing_Well.pdf

[5]Chris Mooney. "Turns Out Wind And Solar Have a Secret Friend: Natural Gas." The Washington Post. August 11, 2016. https://www.washingtonpost.com/news/energy-environment/wp/2016/08/11/turns-out-wind-and-solar-have-a-secret-friend-natural-gas

[6]"The Secret to Côte D'ivoire's Electric Success." The World Bank. July 23, 2020. https://www.worldbank.org/en/news/feature/2020/07/23/the-secret-to-cote-divoires-electric-success

[7]Paul Burkhardt. "Coal-Reliant South Africa Is Turning to Gas Power." Bloomberg. August 2, 2021. https://www.bloomberg.com/news/articles/2021-08-03/coal-reliant-south-africa-invests-in-gas-power-to-go-greener

[8]Brian Gicheru Kinyua. "Equatorial Guinea Successfully Develops Offshore Gas 'Mega Hub'." March 10, 2021. https://www.maritime-executive.com/article/equatorial-guinea-successfully-develops-offshore-gas-mega-hub

[9]"Noble Energy Announces Agreement to Progress Development of Alen Natural Gas, Offshore Equatorial Guinea." GlobeNewswire. May 10, 2018. https://www.globenewswire.com/news-release/2018/05/10/1500324/35227/en/Noble-Energy-Announces-Agreement-to-Progress-Development-of-Alen-Natural-Gas-Offshore-Equatorial-Guinea.html

[10]Umesh Ellichipuram. "First Gas Achieved From Noble Energy's Alen Gas Monetization Project." Offshore Technology. November 26, 2021. https://www.offshore-technology.com/news/first-gas-noble-energys-alen-gas-monetization-project

[11]Amina Ali. "South Africa Should Urgently Move on Gas Monetisation to Solve its Energy Problems." African Press Organization. May 19, 2021. https://www.einnews.com/pr_news/541562880/south-africa-should-urgently-move-on-gas-monetisation-to-solve-its-energy-problems-by-amina-ali

[12]Amina Ali. "South Africa Should Urgently Move on Gas Monetisation to Solve its Energy Problems." African Press Organization. May 19, 2021. https://www.einnews.com/pr_news/541562880/south-africa-should-urgently-move-on-gas-monetisation-to-solve-its-energy-problems-by-amina-ali

[13]Ciaran Ryan. "Total's Second Major Gas Find Puts SA on the Global Energy Map." The Writer's Room. November 11, 2020. http://www.writersroom.co.za/totals-second-major-gas-find-puts-sa-on-the-global-energy-map

[14]"Compressed Natural Gas Division." NIPCO PLC. n.d. https://nipcoplc.com/cng-division

[15]Muyiwa Lucas. "'NIPCO's gas investment commendable'." The Nation. April 30, 2021. https://thenationonlineng.net/nipcos-gas-investment-commendable

[16]"Case Studies." Victoria Oil & Gas PLC. n.d. https://www.victoriaoilandgas.com/csr/case-studies

[17]"Cameroon: SNH now owns 5% in Logbaba gas project, alongside British Victoria Oil & Gas." Business in Cameroon. July 5, 2017. https://www.businessincameroon.com/hydrocarbon/0507-7241-cameroon-snh-now-owns-5-in-logbaba-gas-project-alongside-british-victoria-oil-gas

[18]Olushola Bello. "Nigeria: Dangote Investments Generate Employment Opportunities." AllAfrica. May 25, 2021. https://allafrica.com/stories/202105250182.html

[19]"How Many Jobs Can Dangote Refinery Actually Create?" Arbiterz. November 16, 2020. https://arbiterz.com/how-many-jobs-can-dangote-refinery-actually-create

[20]Kirk Chisholm. "144 Products Made From Petroleum and 4 That May Shock You." Innovative Advisory Group. n.d. https://innovativewealth.com/inflation-monitor/what-products-made-from-petroleum-outside-of-gasoline

[21]"The Future of Petrochemicals: Growth Surrounded by Uncertainty." Deloitte. 2019. https://www2.deloitte.com/content/dam/Deloitte/us/Documents/energy-resources/us-the-future-of-petrochemicals.pdf

[22]Lutz Goedde, Amandla Ooko-Ombaka, and Gillian Pais. "Winning in Africa's Agricultural Market." McKinsey & Company. February 15, 2019. https://www.mckinsey.com/industries/agriculture/our-insights/winning-in-africas-agricultural-market

[23]Gordon Cope. "Progress on the Horizon." World Fertilizer. March 10, 2020. https://www.worldfertilizer.com/special-reports/10032020/progress-on-the-horizon

[24]Prince Okafor. "FG Commits $670m on $3.5bn Brass Methanol Plant Construction." Vanguard. January 30, 2021. https://www.vanguardngr.com/2021/01/fg-commits-670m-on-3-5bn-brass-methanol-plant-construction

[25]"Nigeria's BFPCL to Establish Methanol Plant Project in Odioma." International Finance. February 3, 2021. https://internationalfinance.com/nigerias-bfpcl-establish-methanol-plant-project-odioma

[26]"Small Scale LNG." SNAM. June 15, 2021. https://www.snam.it/en/energy_transition/sustainable_mobility/small_scale_lng

[27]Srinivasa Pachipulusu. "Small-Scale LNG: A Reality Today May Be a Game-Changer for Tomorrow." Gas Processing & LNG. n.g. http://www.gasprocessingnews.com/features/201808/small-scale-lng-a-reality-today-may-be-a-game-changer-for-tomorrow.aspx

[28]Mirza Duran. "World's First Converted FLNG in Milestone Loading." Offshore Energy. July 13, 2020. https://www.offshore-energy.biz/worlds-first-converted-flng-in-milestone-loading

[29]Amanda Battersby. "Floating LNG: Agreement To Boost Utilisation for Golar Vessel Offshore Cameroon." Upstream. July 22, 2021. https://www.upstream

online.com/lng/floating-lng-agreement-to-boost-utilisation-for-golar-vessel-offshore-cameroon/2-1-1042636

[30]"What Are the Environmental Benefits of Liquified Natural Gas?" Matrix NAC. 2020. https://www.matrixnac.com/what-are-the-environmental-benefits-of-liquified-natural-gas-lng

[31]Vincent Xia. "Stanford Scientists Creating Ways to Quickly, Accurately and Inexpensively Find Natural Gas Leaks." Stanfords News. June 7, 2019. https://news.stanford.edu/2019/06/07/trying-stop-natural-gas-leaks-wake-destruction

[32]Jonathan Mingle. "Can a New Wave of Laser and Aerial Imagery Technologies Slash Methane Emissions?" GreenBiz. December 9, 2019. https://www.greenbiz.com/article/can-new-wave-laser-and-aerial-imagery-technologies-slash-methane-emissions

[33]Foreign Relations. May 21, 2021. https://www.cfr.org/in-brief/how-cutting-methane-emissions-can-move-needle-climate-change

[34]Matt McGrath. "Climate Change: Curbing Methane Emissions Will 'Buy Us Time'." BBC. August 11, 2021. https://www.bbc.com/news/science-environment-58174111

[35]Karin Rives. "Global Methane Releases Must Come Down 40%-45% by 2030, UN Report Finds." S&P Global. May 6, 2021. https://www.spglobal.com/marketintelligence/en/news-insights/latest-news-headlines/global-methane-releases-must-come-down-40-45-by-2030-un-report-finds-64141685

[36]"Global Assessment: Urgent steps Must Be Taken To Reduce Methane Emissions This Decade." UN Environment Programme. May 6, 2021. https://www.unep.org/news-and-stories/press-release/global-assessment-urgent-steps-must-be-taken-reduce-methane

[37]"Five Green Energy Trends Topping Miners' Agendas in 2021 – Report." MINING.com. January 27, 2021. https://www.mining.com/climate-to-top-energy-and-mining-agendas-in-2021-report/

[38]"Zero Flaring by 2030." The World Bank. n.d. https://www.worldbank.org/en/programs/zero-routine-flaring-by-2030#4

[39]Jonathan Mingle. "Can a New Wave of Laser and Aerial Imagery Technologies Slash Methane Emissions?" GreenBiz. December 9, 2019. https://www.greenbiz.com/article/can-new-wave-laser-and-aerial-imagery-technologies-slash-methane-emissions

[40]Cheryl Katz. "Satellites Seek Out Methane Leaks From Pipelines, Oil Fields, Landfills and Farms." PBS. June 17, 2021. https://www.pbs.org/newshour/science/satellites-seek-out-methane-leaks-from-pipelines-oil-fields-landfills-and-farms

[41]Megan Horn. "Using Drone Technology in the Fight Against Methane Emissions." Innovate Energy. February 15, 2021. https://innovateenergynow.com/resources/drone-methane-detection

Chapter 7: Hydrogen Can Power Tremendous Growth for Africa—If We're Smart About It

[1]"South Africa Hydrogen Valley Final Report." Department of Science and Innovation, Republic of South Africa. October 2021. https://www.dst.gov.za/images/2021/Hydrogen_Valley_Feasibility_Study_Report_Final_Version.pdf

[2]"South Africa's Massive Hydrogen Valley Project Is a Step Closer – Here's What You Need To Know." BusinessTech. October 12, 2021. https://businesstech.co.za/news/energy/528408/south-africas-massive-hydrogen-valley-project-is-a-step-closer-heres-what-you-need-to-know

[3]Jonathan Gaventa. "Will the Dash for Hydrogen Benefit Sub-Saharan Africa?" Energy Monitor. November 5, 2021. https://www.energymonitor.ai/tech/hydrogen/will-the-dash-for-hydrogen-benefit-sub-saharan-africa

[4]"An Investor's Guide to Hydropower in Africa." International Water Power & Dam Construction. December 3, 2021. https://www.waterpowermagazine.com/features/featurean-investors-guide-to-hydropower-in-africa-9297435/#:~:text=Africa%20also%20has%20the%20highest,North%20America%2C%20Asia%20and%20Europe

[5]Nick van Mead. "Fuel of the Future: How To Realize the Potential of Hydrogen." World Economic Forum. March 7, 2022. https://www.weforum.org/agenda/2022/03/hydrogen-decarbonization-climate-change-energy

[6]Bernd Radowitz. "Renewables and Green Hydrogen in Europe Will Accelerate in Response to Russia's War: DNV." Recharge. April 8, 2022. www.rechargenews.com/energy-transition/renewables-and-green-hydrogen-in-europe-will-accelerate-in-response-to-russias-war-dnv/2-1-1198641

[7]Bernd Radowitz. "Germany Eyes World's Cheapest Green Hydrogen From Namibia Amid Global 'Race for Best Sites'." Recharge. August 25, 2021. https://www.rechargenews.com/energy-transition/germany-eyes-worlds-cheapest-green-hydrogen-from-namibia-amid-global-race-for-best-sites/2-1-1057335

[8]Hans-Joachim Rickel. "West Africa Can Become the Climate-Friendly Energy Powerhouse of the World." Federal Ministry of Education and Research, Germany. May 27, 2021. https://www.bmbf.de/bmbf/en/home/_documents/west-africa-can-become-the-cli-energy-powerhouse-of-the-world.html#:~:text=The%20hydrogen%20potential%20atlas%20will,is%20suitable%20for%20wind%20turbines.

[9]"60 "Masters" for Green Hydrogen From West Africa." H2 Atlas Africa. January 21, 2022. https://www.h2atlas.de/en/news/detail/Kick-off-Germany-of-IMP-EGH

[10]Matthew Goosen. "Germany to Invest Over $45 Million in South Africa's Green Hydrogen Economy." Energy Capital & Power. n.d. https://energycapitalpower.com/germany-to-invest-over-45-million-in-south-africas-green-hydrogen-economy

[11]Bernd Radowitz. "Germany Eyes World's Cheapest Green Hydrogen From Namibia Amid Global 'Race for Best Sites.'" Recharge. August 25, 2021. https://www.rechargenews.com/energy-transition/germany-eyes-worlds-cheapest-green-hydrogen-from-namibia-amid-global-race-for-best-sites/2-1-1057335

[12]Philippe Benoit. "Unlocking Green Energy in Africa Can Impact Climate Change Globally." The Hill. June 9, 2021. https://thehill.com/opinion/energy-environment/557599-unlocking-green-energy-in-africa-can-impact-climate-change

[13]Iain Esau. "'It's Time To Think Big' Says Merkel's Man in Africa." Upstream. December 13, 2021. https://www.upstreamonline.com/energy-transition/its-time-to-think-big-says-merkels-man-in-africa/2-1-1107376

[14]Philippe Benoit. "To Fund Grand Inga Using Green Hydrogen, Equity and Ethics Matter." Inter Press Service. June 18, 2021. https://www.ipsnews.net/2021/06/fund-grand-inga-using-green-hydrogen-equity-ethics-matter

[15]Leigh Collins. "British-German Developer Wins Tender To Build 3GW Green Hydrogen Project in Namibia." Recharge. November 5, 2021. https://www.rechargenews.com/energy-transition/british-german-developer-wins-tender-to-build-3gw-green-hydrogen-project-in-namibia/2-1-1094767

[16]"Belgium-Namibia MoU on Green Hydrogen." International Energy Agency. February 11, 2022. https://www.iea.org/policies/14754-belgium-namibia-mou-on-green-hydrogen

[17]"Namibian Port signs MoU with Port of Rotterdam, to Become Green Hydrogen Export Hub for Europe and the Rest of the World." Hydrogen Central. November 15, 2021. https://hydrogen-central.com/namibian-port-port-rotterdam-green-hydrogen-export-hub-europe/#:~:text=5%20min%20read-,Namibian%20Port%20signs%20MoU%20with%20Port%20of%20Rotterdam%2C%20to%20become,and%20distribution%20of%20green%20hydrogen.

[18]"Fuel Cell Electric Vehicles." US Department of Energy Alternative Fuels Data Center. n.d. https://afdc.energy.gov/vehicles/fuel_cell.html

[19]Aaron Denman, Søren Konnerup, Peter Meijer, and Brian Murphy. "Business Opportunities in Low-Carbon Hydrogen." Bain & Company. June 22, 2021. https://www.bain.com/insights/low-carbon-hydrogen-enr-report-2021

[20]Kartik Jayaram, Adam Kendall, Ken Somers, and Lyes Bouchene. "Africa's Green Manufacturing Crossroads." McKinsey & Company. 27 September 2021. https://www.mckinsey.com/~/media/mckinsey/business%20functions/sustainability/our%20insights/africas%20green%20manufacturing%20crossroads/africas-green-manufacturing-crossroads-choices-for-a-low-carbon-industrial-future.pdf

Chapter 8: Solar Power and Wind: More Work To Be Done

[1]Zhenya Liu. "Global Energy Interconnection." China Electric Power Press published by Elsevier: Boston. 2015. p. 30.

[2]"Electricity Information: Overview | Electricity Production." International Energy Agency. 2021. https://www.iea.org/reports/electricity-information-overview/electricity-production

[3]Linda Munyengeterwa and Sean Whittaker. "Powering Africa's Sustainable Development Through Wind." World Bank Blogs. June 24, 2021. https://blogs.worldbank.org/climatechange/powering-africas-sustainable-development-through-wind

[4]"Barriers to Renewable Energy Technologies." Union of Concerned Scientists. December 20, 2017. https://www.ucsusa.org/resources/barriers-renewable-energy-technologies

[5]"Lazard's Levelized Cost of Energy Analysis." Lazard. 2017. https://www.lazard.com/media/450337/lazard-levelized-cost-of-energy-version-110.pdf

[6]Michael J. Coren. "The Era of Subsidies for Wind and Solar May Be Ending Far Too Soon." Yahoo Life. April 17, 2021. https://www.yahoo.com/lifestyle/era-subsidies-wind-solar-may-120048344.html

[7]Michael Taylor. "Energy Subsidies: Evolution in the Global Energy Transformation to 2050." International Renewable Energy Agency. January 2020. https://www.irena.org/-/media/Files/IRENA/Agency/Publication/2020/Apr/IRENA_Energy_subsidies_2020.pdf

[8]"Statement from OECD Secretary-General Mathias Cormann on Climate Finance in 2019." Organisation for EconomicCo-operation and Development. September 17, 2021. https://www.oecd.org/newsroom/statement-from-oecd-secretary-general-mathias-cormann-on-climate-finance-in-2019.htm

[9]"Global Renewables Outlook: Energy Transformation 2050." International Renewable Energy Agency. 2020. https://www.irena.org/-/media/Files/IRENA/Agency/Publication/2020/Apr/IRENA_Global_Renewables_Outlook_2020.pdf

[10]Rebekah Shirley. "Bridging the Skills Gap Is Key for Energy Access, New Jobs." World Bank Blogs. November 12, 2018. https://blogs.worldbank.org/energy/bridging-skills-gap-key-energy-access-new-jobs

[11]"Exploring the Gap Between Skills and the Renewable Energy Industry." AltGen. 2022. https://www.altgen.com/2021/05/10/exploring-the-gap-between-skills-and-the-renewable-energy-industry

[12]"Energy Efficiency." Bonneville Power Administration. n.d. https://www.bpa.gov/EE/Technology/demand-response/Pages/intermittent-renewable-energy.aspx

[13]Fred Lambert. "Tesla Reveals Megapack Prices: Starts at $1 Million." Electrek July 26, 2021. https://electrek.co/2021/07/26/tesla-reveals-megapack-prices

[14]Tariye Gbadegesin. "Africa Must Assume its Place in the Global Battery Race." The Africa Report. January 7, 2020. https://www.theafricareport.com/21865/africa-must-assume-its-place-in-the-global-battery-race

[15]Halim Nazar. "Chinese Mining in the DRC: From Sicomines to Global Cobalt Monopoly." The Institute of Chinese Studies. August 27, 2021. https://icsin.org/blogs/2021/08/27/chinese-mining-in-the-drc-from-sicomines-to-global-cobalt-monopoly/#:~:text=Under%20the%20agreement%2C%20China%20would,infrastructure%20and%20mining%20credits%20combined

[16]"The Toll of the Cobalt Mining Industry on Health and the Environment. CBS News. March 6, 2018. https://www.cbsnews.com/news/the-toll-of-the-cobalt-mining-industry-congo

[17]"This Is What We Die For: Human Rights Abuses in the Democratic Republic of the Congo Power the Global Trade in Cobalt." Amnesty International. 2016. https://www.amnesty.org/en/documents/afr62/3183/2016/en

[18]David Lawrence. "Investors Forecast Bright Future for Mini-Grids in Africa." International Finance Corporation. June 2020. https://www.ifc.org/wps/wcm/connect/news_ext_content/ifc_external_corporate_site/news+and+events/news/insights/africa-mini-grids

[19]Jonathan Phillips, Benjamin Attia, and Victoria Plutshack. "Lessons From the Proliferating Mini-Grid Incentive Programs in Africa. Brookings. December 11, 2020. https://www.brookings.edu/blog/future-development/2020/12/11/lessons-from-the-proliferating-mini-grid-incentive-programs-in-africa

[20]Ruchi Soni and Takehiro Kawahara. "Three Key Challenges to Scale up the Mini-Grid Sector." Sustainable Energy for All. July 23, 2020. https://www.seforall.org/news/three-key-challenges-to-scale-up-the-mini-grid-sector

[21]Ruchi Soni and Takehiro Kawahara. "Three Key Challenges to Scale up the Mini-Grid Sector." Sustainable Energy for All. July 23, 2020. https://www.seforall.org/news/three-key-challenges-to-scale-up-the-mini-grid-sector

[22]David Lawrence. "Investors Forecast Bright Future for Mini-Grids in Africa." International Finance Corporation. June 2020. https://www.ifc.org/wps/wcm/connect/news_ext_content/ifc_external_corporate_site/news+and+events/news/insights/africa-mini-grids

[23]Feyza Sahinyazan and Serasu Duran. "Why Renewable Energy 'Mini-Grids' in Remote Communities Fail and How To Avoid it." The Conversation. August 3, 2021. https://theconversation.com/why-renewable-energy-mini-grids-in-remote-communities-fail-and-how-to-avoid-it-164231

[24]Pier Paolo Raimondi. "The Scramble for Africa's Rare Earths: China Is not Alone. Italian Institute for International Political Studies. June 7, 2021. https://www.ispionline.it/en/publication/scramble-africas-rare-earths-china-not-alone-30725

[25]M. Garside. "Major Countries in Copper Mine Production Worldwide in 2021." Statista. March 11, 2022. https://www.statista.com/statistics/264626/copper-production-by-country

[26]Sanusi Alhaji Mohammed. "Steel and Trade in Sub-Saharan Africa." 4th Regional Meeting on Trade in Steel Products. June 5, 2020. https://www.mesteel.com/lecture/steelinafrica.htm

Chapter 9: Financing a Just Energy Transition: Our Path to Overcoming Obstacles, Interference, and Unkept Promises

[1]Kate Linebaugh. "The Fight Over Climate Change's Price Tag." The Journal. November 4, 2021. https://www.wsj.com/podcasts/the-journal/the-fight-over-climate-change-price-tag/bb8dada0-18db-49a1-91a6-f7c8af33a4cb

[2]"Africa Energy Review 2021." PricewaterhouseCoopers. November 2021. https://www.pwc.com/ng/en/assets/pdf/africa-energy-review-2021.pdf

[3]"Step up Climate Change Adaptation or Face Serious Human and Economic Damage – UN Report." UN Environment Programme. January 14, 2021. https://www.unep.org/news-and-stories/press-release/step-climate-change-adaptation-or-face-serious-human-and-economic

[4]"Energy Transition Investment Hit $500 Billion in 2020 – For First Time." BloombergNEF. January 19, 2021. https://about.bnef.com/blog/energy-transition-investment-hit-500-billion-in-2020-for-first-timel

[5]Jocelyn Timperley. "The Broken $100-Billion Promise of Climate Finance - and How to Fix it." Nature. October 20, 2021. https://www.nature.com/articles/d41586-021-02846-3

[6]Somini Sengupta. "The Rich World's Promise of $100 Billion in Climate Aid Inches Forward." New York Times. October 25, 2021. https://www.nytimes.com/2021/10/25/climate/100-billion-climate-aid-cop26.html

[7]Jocelyn Timperley. "The Broken $100-Billion Promise of Climate Finance - and How to Fix it." Nature. October 20, 2021. https://www.nature.com/articles/d41586-021-02846-3

[8]Kingsley Jeremiah. "Why Africa's $6.7tr Fossil Fuels Resources May Remain Stranded." The Guardian. November 11, 2021. https://guardian.ng/business-services/business/why-africas-6-7tr-fossil-fuels-resources-may-remain-stranded

[9]"UN List of Least Developed Countries." United Nations Conference on Trade and Development. n.d. https://unctad.org/topic/least-developed-countries/list

[10]Jocelyn Timperley. "The Broken $100-Billion Promise of Climate Finance - and How to Fix it." Nature. October 20, 2021. https://www.nature.com/articles/d41586-021-02846-3

[11]Nkiruka Nnorom. "Africa Requires $2.8trn Investment to Transition to Clean Energy by 2050 – PWC." Vanguard. November 8, 2021. https://www.vanguardngr.com/2021/11/africa-requires-2-8trn-investment-to-transition-to-clean-energy-by-2050-pwc

[12]"Scaling Up Renewable Energy Deployment in Africa." International Renewable Energy Agency. January 2019. https://irena.org/-/media/Files/IRENA/Agency/Publication/2019/Jan/IRENA_Africa_impact_2019.pdf?la=en&hash=6B16ABE754FF6F843601E1E362F5D6B730ADF7A2

[13]Jeff Mason, Andrea Shalal, and Emma Rumney. "South Africa To Get $8.5 bln From US, EU and UK to Speed up Shift From Coal." Reuters.

November 2, 2021. https://www.reuters.com/business/environment/us-eu-others-will-invest-speed-safricas-transition-clean-energy-biden-2021-11-02

[14]"Renewable Energy Sector Attracts Foreign Investors to Africa." De Aar Solar Power. n.d. https://deaarsolar.co.za/renewable-energy-sector-attracts-foreign-investors-africa/#:~:text=%E2%80%9CConsidering%20that%20South%20Africa%20has,Pickering%2C%20Managing%20Director%20of%20Globeleq

[15]Natasha Odendaal. "South African Investors Bulk Financiers of Renewable Energy Success." Creamer Media. October 24, 2014. https://www.engineeringnews.co.za/article/south-african-investors-bulk-financiers-of-renewable-energy-success-2014-10-24/rep_id:4136

[16]Luzuko Nomjana. "REIPPP Comes of Age." FutureGrowwth. February 18, 2020. https://futuregrowth.co.za/insights/reippp-comes-of-age

[17]Terence Creamer. "Govt Signals Reduction or Elimination of Guarantees for REIPPPP Projects." Creamer Media. February 23, 2022. https://www.engineeringnews.co.za/article/govt-signals-reduction-or-elimination-of-guarantees-for-reipppp-projects-2022-02-23

[18]James Gavin. "Financing Africa's Energy Transition." African Business. August 25, 2020. https://african.business/2020/08/energy-resources/financing-africas-energy-transition

[19]James Gavin. "Financing Africa's Energy Transition." African Business. August 25, 2020. https://african.business/2020/08/energy-resources/financing-africas-energy-transition

[20]"29th Afreximbank Annual Meetings." Afreximbank. June 2022. https://www.afreximbank.com

[21]Xolisa Phillip. "Africa: 'No Shame in Funding Fossil Fuel Projects' Says Afreximbank President." The Africa Report. November 18, 2021. https://www.theafricareport.com/147364/africa-no-shame-in-funding-fossil-fuel-projects-says-afreximbank-president

[22]"Afreximbank, African Petroleum Producers Organization (APPO) Sign MoU for the Creation of a Multi-Billion-Dollar Energy Bank." Yahoo Finance. May 17,

2022. https://finance.yahoo.com/news/afreximbank-african-petroleum-producers-organization-131500007.html

[23]Carl Manlan. "It's Time Wealth - Not Poverty - Dictates Africa's development." World Economic Forum. January 8, 2019. https://www.weforum.org/agenda/2019/01/africa-s-philanthropic-potential

[24]"The State of African Energy 2022." African Energy Chamber. 2022. https://energychamber.org/wp-content/uploads/AEC-Outlook-2022_b.pdf

[25]"African Pension Funds Have Grown Impressively." The Economist. October 2, 2021. https://www.economist.com/middle-east-and-africa/2021/10/02/african-pension-funds-have-grown-impressively

[26]African Energy Chamber. "Demand, Supply and Price Trends: The African Energy Chamber (AEC) Investigates Global Gas Trends in its Q1 2022 Outlook." APO Group Africa Newsroom. February 24, 2022. https://www.africa-newsroom.com/press/demand-supply-and-price-trends-the-african-energy-chamber-aec-investigates-global-gas-trends-in-its-q1-2022-outlook?lang=en

Chapter 10: Africa's Oil and Gas Industry Needs Good Governance More Than Ever

[1]Aaron Sayne. "Nigeria's Petroleum Industry Bill: Improving Sector Performance Through Strong Transparency and Accountability Provisions." Facility for Oil Sector Transparency in Nigeria. July 2011. https://resourcegovernance.org/sites/default/files/documents/nigerias_pib.pdf

[2]"HB 159." https://resourcegovernance.org/sites/default/files/documents/nigerias_pib.pdf

[3]"Nigerian President Signs Long-Delayed Oil Bill into Law." Yahoo News. August 16, 2021. https://news.yahoo.com/nigerian-president-signs-long-delayed-142647779.html

[4]"The Petroleum Industry Bill (PIB): Top 20 Changes You Should Know!" PWC Nigeria Tax Blog. July 7, 2021. https://pwcnigeria.typepad.com/tax_matters_nigeria/2021/07/the-petroleum-industry-bill-pib-top-20-changes-you-should-know.html

[5]"World Bank Says Guyana Economy a Standout in the Hemisphere, But…" The World Bank. April 29, 2022. https://www.stabroeknews.com/2022/04/29/business/world-bank-says-guyana-economy-a-standout-in-the-hemisphere-but

[6]"Assala Energy Invests $24 Million in Export Infrastructure in Gabon." Energy Capital & Power. October 21, 2020. https://energycapitalpower.com/assala__energy_invests__24-million-in-export-infrastructure-in-gabon

[7]"Mozambique: Why 2021 Is the Best Year to Invest in the Country." IN-VR. 2021. https://www.in-vr.co/post/mozambique-why-2020-is-the-best-year-to-invest-in-the-country

[8]"CPI 2021 for Sub-Saharan Africa: Amid Democratic Turbulence, Deep-Seated Corruption Exacerbates Threats to Freedoms." Transparency International. January 25, 2022. https://www.transparency.org/en/news/cpi-2021-sub-saharan-africa-amid-democratic-turbulence-deep-seated-corruption

Chapter 11: Our Governments Must Lay a Path for a Successful Energy Transition

[1]"Yaoundé IV, Cameroon." Ren21. 2021. https://www.ren21.net/cities-2021/cities/yaounde/yaounde

[2]"Cocody, Côte D'Ivoire." Ren21. 2021. https://www.ren21.net/cities-2021/cities/cocody/cocody

[3]"Kampala, Uganda." Ren21. 2021. https://www.ren21.net/cities-2021/cities/kampala/kampala

[4]"Kenya Green Economy Strategy and Implementation Plan 2016-2030." Green Policy Platform. August 2016. https://www.greengrowthknowledge.org/national-documents/kenya-green-economy-strategy-and-implementation-plan-2016-2030

[5]"Kenya Green Economy Strategy and Implementation Plan 2016-2030." Green Policy Platform. August 2016. https://www.greengrowthknowledge

.org/national-documents/kenya-green-economy-strategy-and-implementation-plan-2016-2030

[6]"Skills for Green Jobs in Ghana." International Labour Organization. 2018. https://www.ilo.org/wcmsp5/groups/public/---ed_emp/---ifp_skills/documents/publication/wcms_706868.pdf

[7]Anne Njeri Njoroge. "Green Jobs Creation in Kenya." Strathmore University Energy Research Centre. August 26, 2021. https://serc.strathmore.edu/green-jobs-creation-in-kenya

[8]Ameenah Gurib-Fakim and Landry Signé. "Investment in Science and Technology Is Key to an African Economic Boom." Brookings. January 26, 2022. https://www.brookings.edu/blog/africa-in-focus/2022/01/26/investment-in-science-and-technology-is-key-to-an-african-economic-boom

[9]"Open Africa Power: The Renewable Energy Training Program." Enel. January 29, 2019. https://www.enelgreenpower.com/media/news/2019/01/open-africa-power-the-renewable-energy-training-program-for-the-development-of-the-continent

[10]"Open Africa Power 2021: Call for Applications." Enel. March 4, 2021. https://www.enelfoundation.org/all-news/news/2021/03/open-africa-power-2021--call-for-applications

[11]"Open Africa Power, Told by the 2022 Students." Enel. May 25, 2022. https://www.enelfoundation.org/all-news/news/2022/05/open-africa-power--told-by-the-2022-students

[12]Raad Alkadiri and Björn Ewers. "Preparing National Oil Companies for a New Energy Landscape." Boston Consulting Group. November 23, 2020. https://www.bcg.com/publications/2020/strategies-for-being-an-environmentally-conscious-oil-company

[13]Tariye Gbadegesin. "Africa Must Assume its Place in the Global Battery Race." The Africa Report. January 7, 2020. https://www.thenationalnews.com/business/rising-electric-vehicle-demand-puts-africa-on-the-cusp-of-a-mining-boom-1.1162346; https://www.theafricareport.com/21865/africa-must-assume-its-place-in-the-global-battery-race

[14]Franck Kuwonu. "Africa's Free Trade Area Opens for Business." Africa Renewal. January 7, 2021. https://www.un.org/africarenewal/magazine/january-2021/afcfta-africa-now-open-business

[15]"COP 25: 'Africa's Future Depends on Solidarity': Leaders and Development Partners Rally Around Climate Change Goals." African Development Bank. December 12, 2019. https://www.afdb.org/en/news-and-events/press-releases/cop-25-africas-future-depends-solidarity-leaders-and-development-partners-rally-around-climate-change-goals-33200

Chapter 12: OPEC—Yes, OPEC—Makes Some Great Points About Energy and Climate Concerns

[1]Sam Meredith. "'Our Biggest Compliment Yet': Greta Thunberg Thanks OPEC for Criticism." CNBC. July 5, 2019. https://www.cnbc.com/2019/07/05/greta-thunberg-thanks-oil-cartel-opec-for-climate-change-criticism.html

[2]Mohammad Sanusi Barkindo. "OPEC Statement to the UN Climate Change Conference." Organization of the Petroleum Exporting Countries. December 10, 2019. https://www.opec.org/opec_web/en/press_room/5809.htm

[3]"Enabling Large-Scale CCS Using Offshore CO2 Utilization and Storage Infrastructure Developments." Carbon Sequestration Leadership Forum. November 8, 2017. https://www.cslforum.org/cslf/sites/default/files/documents/7thMinUAE2017/Offshore_CO2- EOR_Final_02_Dec_2017.pdf

[4]Suraje Dessai. "An Analysis of the Role of OPEC as a G77 Member at the UNFCCC." World Wildlife Fund Inc. December 2004. https://wwfeu.awsassets.panda.org/downloads/opecfullreportpublic.pdf

[5]"UN Climate Chief Calls on Oil Producers to Undertake Rapid, Ambitious Energy Transition." United Nations Climate Change. June 20, 2018. https://unfccc.int/news/un-climate-chief-calls-on-oil-producers-to-undertake-rapid-ambitious-energy-transition

[6]Andreas Exarheas. "OPEC Holds Climate Change Workshop." RigZone. October 4, 2020. https://www.rigzone.com/news/opec_holds_climate_change_workshop-04-oct-2020-163461-article

[7]Tina Casey. "UAE Invests in Green Hydrogen for Zero-Emission Aircraft." Triple Pundit. January 22, 2021. https://www.triplepundit.com/story/2021/uae-green-hydrogen/717706

[8]Ivana Kottasová and Mohammed Tawfeeq. "Oil-Rich UAE Opens the Arab World's First Nuclear Power Plant. Experts Question Why." CNN. August 1, 2020. https://edition.cnn.com/2020/08/01/middleeast/nuclear-power-uae-intl/index.html

[9]Dania Saadi. "UAE Juggling Energy Transition, Eking Out Profit From Oil and Gas." S&P Global Commodity Insights. January 18, 2021. https://www.spglobal.com/platts/en/market-insights/latest-news/electric-power/011821-feature-uae-juggling-energy-transition-eking-out-profit-from-oil-and-gas

[10]Inès Magoum. "Equatorial Guinea: BDEAC Lends €122 Million for Sendje Hydropower Plant." Afrik21. June 30, 2020. https://www.afrik21.africa/en/equatorial-guinea-bdeac-lends-e122-million-for-sendje-hydropower-plant

[11]Darrell Proctor. "Algeria Targets Renewables to Diversify Generation." POWER. July 1, 2020. https://www.powermag.com/algeria-targets-renewables-to-diversify-generation

[12]Mohammad Sanusi Barkindo. "Keynote address by OPEC Secretary General." Organization of the Petroleum Exporting Countries. February 26, 2020. https://www.opec.org/opec_web/en/press_room/5854.htm

[13]"Energy in Côte d'Ivoire." The OPEC Fund for International Development. May 6, 2020. https://opecfund.org/news/energy-in-cote-d-ivoire

[14]Justine Würtz. "Combating Energy Poverty in Mali." The OPEC Fund for International Development. April 1, 2017. https://opecfund.org/news/combating-energy-poverty-in-mali

[15]Silvia Mateyka. "The OPEC Fund and Renewable Energy." The OPEC Fund for International Development. February 3, 2020. https://opecfund.org/news/energy-in-cote-d-ivoire

[16]Mohammad Sanusi Barkindo. "Keynote Address." Organization of the Petroleum Exporting Countries. February 26, 2020. https://www.opec.org/opec_web/en/press_room/5854.htm

Chapter 13: Creating Opportunities for Women Empowers All Africans

[1]"Seven Women Entrepreneurs of Solar Energy." International Energy Agency. November 2019. https://www.iea.org/reports/seven-women-entrepreneurs-of-solar-energy

[2]Chloe Sorvino. "Why This 48-Year-Old Woman Is Building Ghana's Biggest Solar Farm. Forbes. Jul 31, 2018. https://www.forbes.com/sites/chloesorvino/2018/07/31/ghana-solar-farm-ubi-salma-okonkwo/?sh=1b43f4271d24

[3]"Renewable Energy: A Gender Perspective." International Renewable Energy Agency. January 2019. https://www.irena.org/publications/2019/Jan/Renewable-Energy-A-Gender-Perspective

[4]"Facts and Figures: Economic Empowerment." UN Women. July 2018. https://www.unwomen.org/en/what-we-do/economic-empowerment/facts-and-figures

[5]Marcus Noland, Tyler Moran, and Barbara Kotschwar. "Is Gender Diversity Profitable? Evidence from a Global Survey." Peterson Institute for International Economics. February 2016. https://www.piie.com/publications/working-papers/gender-diversity-profitable-evidence-global-survey

[6]Jessica Zartler. "A Simple Change Could Give You a 63 Percent Higher Return on Investment." Forbes. August 27, 2016. https://www.forbes.com/sites/womensmedia/2016/08/27/a-simple-change-could-give-you-a-63-percent-higher-return-on-investment/?sh=37f31a88769d

[7]"Nigeria's Laws Hold Women Back, and the Economy Suffers." Council on Foreign Relations. March 8, 2019. https://www.cfr.org/blog/nigerias-laws-hold-women-back-and-economy-suffers

[8]"Tofe Ayeni. "Kenya's Renewable Sector and Women in the Industry: Elizabeth Rogo AEC President East Africa." The Africa Report. May 26, 2020. https://www.theafricareport.com/28520/kenyas-renewable-sector-and-women-in-the-industry-elizabeth-rogo-aec-president-east-africa

[9]"13 Reasons Why Girls Are Not in School on International Day of the Girl Child." United Nations Office for the Coordination of Humanitarian

Affairs. October 11, 2017. https://reliefweb.int/report/world/13-reasons-why-girls-are-not-school-international-day-girl-child

[10]"International Day of Women and Girls in Science: Addressing and Transforming the Gender Gap." WEBWIRE. February 4, 2021. https://www.webwire.com/ViewPressRel.asp?aId=269929

[11]Estelle Marais. "Encouraging Women in Africa to the Forefront of Nuclear Science." International Atomic Energy Agency. September 20, 2018. https://www.iaea.org/newscenter/news/encouraging-women-in-africa-to-the-forefront-of-nuclear-science

[12]"Advancing Gender in the Environment: Making the Case for Women in the Energy Sector." US AID. August 13, 2021. https://www.usaid.gov/sites/default/files/documents/1865/IUCN-USAID-Making_case_women_energy_sector.pdf

[13]"Women Matter: Africa." McKinsey & Company. August 2016. https://www.mckinsey.com/~/media/McKinsey/Featured%20Insights/Women%20matter/Women%20matter%20Africa/Women%20Matter%20Africa%20August%202016.ashx

[14]"Nearly 235 Million Women Worldwide Lack Legal Protections From Sexual Harassment at Work." Social Work Today. n.d. https://www.socialworktoday.com/news/dn_111317.shtml

[15]Monica F. Tejada and Denise Mortimer. "End Gender-Based Violence in the World of Work." Power Africa. November 27, 2018. https://medium.com/power-africa/end-gender-based-violence-in-the-world-of-work-e235c78dc8ca

[16]"Promoting Gender Diversity and Inclusion in the Oil, Gas and Mining Extractive Industries." The Advocates for Human Rights. January 2019. https://unece.org/fileadmin/DAM/energy/images/CMM/CMM_CE/AHR_gender_diversity_report_FINAL.pdf

[17]Monica F. Tejada and Denise Mortimer. "End Gender-Based Violence in the World of Work." Power Africa. November 27, 2018. https://medium.com/power-africa/end-gender-based-violence-in-the-world-of-work-e235c78dc8ca

[18]"Women's Economic Empowerment in Oil and Gas Industries in Africa." African Development Bank African Natural Resources Center. 2017. https://www.afdb.org/fileadmin/uploads/afdb/Documents/Publications/anrc/AfDB_WomenEconomicsEmpowerment_V15.pdf

[19]Challenges Faced by Women Entrepreneurs in Africa." Africa.com. November 6, 2017. https://africa.com/challenges-faced-women-entrepreneurs-africa

[20]"AWEAP Introduction." African Women in Energy and Power (AWEaP). 2022. https://aweap.africa/aweap-introduction

[21]"African Energy Companies Join Equal by 30 Campaign and Commit to Equal Pay, Leadership and Opportunities for Women in Energy." African Energy Chamber. May 7, 2020. https://energychamber.org/african-energy-companies-join-equal-by-30-campaign-and-commit-to-equal-pay-leadership-and-opportunities-for-women-in-energy

Chapter 14: The Value of Cooperation Among Oil and Gas Stakeholders and Environmentalists

[1]Mel Frykberg. "The Future of Africa's Energy Sector Is Big and Bright, Says Pan African Businessman." Independent Online (IOL). October 2, 2018. https://www.iol.co.za/business-report/energy/feature-the-future-of-africas-energy-sector-is-big-and-bright-says-pan-african-businessman-17315849

[2]"Zambia Takes the Keys Away from 'Drivers' of Deforestation." The World Bank. March 2, 2018. https://www.worldbank.org/en/news/feature/2018/03/02/zambia-takes-the-keys-away-from-drivers-of-deforestation

[3]"New Energies From the African Continent." Eni. n.d. https://www.eni.com/en-IT/global-presence/africa.html

[4]Hassan Sachedina and Kerry Hancock. "Biodiversity and Local Communities." Eni. n.d. https://www.eni.com/en-IT/low-carbon/tutela-biodiversita-benefici-comunita.html

[5]"Climate Risk & Resilience Report 2020." Tullow Oil. 2020. https://www.tullowoil.com/application/files/2916/2140/7519/Tullow_Oil_-_Climate_Risk__Resilience_Report_2020_.pdf

[6]"The March 5, 1987, Ecuador Earthquakes: Mass Wasting and Socioeconomic Effects." National Academies of Sciences, Engineering, and Medicine. 1991. https://nap.nationalacademies.org/read/1857/chapter/8#86

[7]"Protecting the Environment in Mozambique's Emerging Oil and Gas Sector." UN Environment Programme. March 18, 2019. https://www.unep.org/news-and-stories/press-release/protecting-environment-mozambiques-emerging-oil-and-gas-sector

[8]George Prpich, Kabari Sam, and Frédéric Coulon. "Stakeholder Engagement and Sustainable Environmental Management of Oil Contaminated Sites in Nigeria." In: Energy in Africa: Policy, Management and Sustainability. August 2018. https://core.ac.uk/download/pdf/188365533.pdf

[9]J. Peter Pham. "Africa's Real Strategic Import for the Green Economy." Atlantic Council. March 29, 2021. https://www.atlanticcouncil.org/blogs/africasource/africas-real-strategic-import-for-the-green-economy

Chapter 15: Using Wisdom From Other Parts of the Globe to Transition Toward Africa's New Energy Mix

[1]"African Legends." Colours of the Rainbow. n.d. https://www.colours-of-the-rainbow.com/african-legends.html

[2]Fergus Green and Ajay Gambhir. "Transitional Assistance Policies for Just, Equitable and Smooth Low-Carbon Transitions: Who, What and How?" Climate Policy. August 28, 2019. https://www.tandfonline.com/doi/abs/10.1080/14693062.2019.1657379?journalCode=tcpo20%20an

[3]"Natural Gas Vision and Strategy: Energy." Government of Alberta. October 2020. https://open.alberta.ca/dataset/988ed6c1-1f17-40b4-ac15-ce5460ba19e2/resource/a7846ac0-a43b-465a-99a5-a5db172286ae/download/energy-getting-alberta-back-to-work-natural-gas-vision-and-strategy-2020.pdf

[4]Srinivasa Rao Patnana. "Energy Transition in India: Moving Towards Self-Reliance and Sustainability." KPMG. November 4, 2020. https://home.kpmg/in/en/home/insights/2020/11/energy-transition-in-india.html

[5]Ajinkya Shrish Kamat, Venkatesh Narayanamurti, and Radhika Khosla. "What the World Can Learn From Clean Energy Transitions in India, China, and Brazil."

The Conversation. October 2, 2020. https://theconversation.com/what-the-world-can-learn-from-clean-energy-transitions-in-india-china-and-brazil-146773

[6]Oludamilare Bode Adewuyia, Mark Kipngetich Kiptooa, Ayodeji Fisayo Afolayanb, Theophilus Amaraac, Oluwatobi Idowu Alawoded, and Tomonobu Senjyua. "Challenges and Prospects of Nigeria's Sustainable Energy Transition With Lessons From Other Countries' Experiences." Energy Reports. November 2020. https://www.sciencedirect.com/science/article/pii/S2352484719308832

[7]"Market to Watch: Taiwan Offshore Wind." Ocean Energy Resources. January 17, 2020. https://ocean-energyresources.com/2020/01/17/market-to-watch-taiwan-offshore-wind

[8]Aisha Salaudeen. "Senegal Launches Wind Power Plant as Part of its Commitment to Renewable Energy." CNN. February 25, 2020. https://www.cnn.com/2020/02/25/africa/west-africas-largest-wind-plant/index.html

[9]"Wind Atlas for South Africa." South African National Energy Development Institute. n.d. http://www.wasaproject.info/index.html

Chapter 16: Successful African Energy Entrepreneurs

[1]"Meet Vere Shaba." Investec/YouTube. September 4, 2020. https://www.youtube.com/watch?v=WHIiKXbjjDM

[2]"Vere Shaba, a South African Entrepreneur Harnessing the Power of Engineering to Minimise the Environmental Effect of Construction." Lioness of Africa. October 22, 2017. https://www.lionessesofafrica.com/blog/startup-story-of-vere-shaba

[3]"Meet Vere Shaba." Investec/YouTube. September 4, 2020. https://www.youtube.com/watch?v=WHIiKXbjjDM

[4]"Profiles of Change: Youth Opportunities in the Clean Energy Sector." Energy and Environment Partnership Trust Fund (EEP Africa). 2019. https://eepafrica.org/wp-content/uploads/2019/12/Youth-opportunities-brief_Digital.pdf

[5]Salah Koubaa. "Renewable Energy Through the Lens of Entrepreneurship Theory: The case of Morocco." Cairn.info. April 13, 2018. https://www.cairn.info/revue-projectique-2017-3-page-69.htm

[6]Akinyi Ochieng and Fadekemi Abiru. "Solar Startups Are Plugging Africa's Energy Gap." TheBeamMagazine. April 9, 2018. https://medium.com/thebeammagazine/solar-start-ups-are-plugging-africas-energy-gap-14e512fa9228

[7]John-Paul Iwuoh. "Solar Energy Is Africa's Next Big Thing! How Entrepreneurs Can Exploit This Lucrative but Untapped Opportunity." SmallStarter. August 1, 2013. https://www.smallstarter.com/browse-ideas/potentials-of-solar-power-business-in-africa

[8]"#TheBRIDGEMeets – Henri Nyakarundi, Founder & CEO of ARED Group." The Bridge. June 2, 2020. http://www.thebridgecareers.rw/blog/thebridgemeets-henri-nyakarundi-the-founder-ceo-of-ared-group

[9]R. Jallow. "Henri Nyakarundi: Powering Africa one Solar Kiosk at a Time." Djellibah. November 6, 2020. https://djellibah.com/en/home/2020/11/6/henri-nyakarundi

[10]"Nthabiseng Mosia, an Entrepreneur Finding Affordable Clean Energy Solutions for Africa by Harnessing the Power of Solar Technology." Lioness of Africa. December 11, 2016. https://www.lionessesofafrica.com/blog/2016/12/11/startup-story-nthabiseng-mosia

[11]Douglas Broom. "These Solar-Power Entrepreneurs Are Lighting up Sierra Leone." How We Made it in Africa. December 9, 2019. https://www.howwemadeitinafrica.com/these-solar-power-entrepreneurs-are-lighting-up-sierra-leone/63980

[12]www.nthabisengmosia.com

[13]"Ugwem Eneyo, H'16." Harambeans. 2022. https://www.harambeans.com/harambeans_at_work/ugwem-eneyo-h16

[14]https://shyftpower.com

[15]Tola Agunbiade. "Solstice Energy Solutions Wants To Help Nigerian Homeowners Manage Their Power Supply Better." TechCabal. September 8, 2016. https://techcabal.com/2016/09/08/solstice-energy-solutions/

[16]www.linkedin.com/in/dave-lello-58138112

[17]www.ekasi.energy

[18]https://www.linkedin.com/in/hannah-kabir

[19]http://creedsenergy.com

[20]"Hannah Kabir – Our Rebrand Nigeria Ambassador." Rebrand Nigeria. January 22, 2020. https://rebrandnigeria.org/2020/01/22/hannah-kabir-our-rebrand-nigeria-ambassador

[21]"Nigerian Women's Perspectives Light the Way in the Renewable Energy Sector." The Rockefeller Foundation. September 10, 2020. https://www.rockefellerfoundation.org/case-study/nigerian-womens-perspectives-light-the-way-in-the-renewable-energy-sector

[22]www.linkedin.com/in/habibaali

[23]"Nigerian Women's Perspectives Light the Way in the Renewable Energy Sector." The Rockefeller Foundation. September 10, 2020. https://www.rockefellerfoundation.org/case-study/nigerian-womens-perspectives-light-the-way-in-the-renewable-energy-sector

[24]Alana Chloe Esposito. "Habiba Ali Brings Renewable Energy to Nigeria." A Woman's Thing. October 25, 2018. https://awomensthing.org/blog/habiba-ali-brings-renewable-energy-to-nigeria

[25]www.linkedin.com/in/opio-derrick-hosea-51632036

[26]"Derrick Hosea Opio." Meaningful Business Limited. 2022. https://meaningful.business/team/derrick-hosea-opio

[27]Kate Douglason. "How Lois Gicheru Is Moving Kenya Towards Green Energy." How We Made it in Africa. September 14, 2012. https://www.howwemadeitinafrica.com/how-lois-gicheru-is-moving-kenya-towards-green-energy/20220

Chapter 17: Energy Development Can Help Solve Africa's "Missing Jobs" Problem

[1]Louise Fox, Philip Mader, James Sumberg, Justin Flynn, and Marjoke Oosterom. "Africa's 'Youth Employment' Crisis Is Actually a 'Missing Jobs' Crisis." Brookings. September 8, 2020. https://www.brookings.edu/research/africas-youth-employment-crisis-is-actually-a-missing-jobs-crisis

[2]"Why Is Guyana Poor?" The Borgen Project. September 17, 2017. https://borgenproject.org/why-is-guyana-poor

[3]"Guyana Is no Longer the Second Poorest Nation in the Western Hemisphere." Kaieteur News. May 4, 2010. https://www.kaieteurnewsonline.com/2010/05/04/guyana-is-no-longer-the-second-poorest-nation-in-the-western-hemisphere

[4]"ExxonMobil Announces Significant Oil Discovery Offshore Guyana." ExxonMobil. May 20, 2015. https://corporate.exxonmobil.com/Locations/Guyana/News-releases/ExxonMobil-announces-significant-oil-discovery-offshore-Guyana

[5]"ExxonMobil says second well offshore Guyana confirms significant oil discovery." ExxonMobile. June 30, 2016. https://corporate.exxonmobil.com/News/Newsroom/News-releases/2016/0630_ExxonMobil-says-second-well-offshore-Guyana-confirms-significant-oil-discovery

[6]"ExxonMobil Announces Oil Discovery at Whiptail, Offshore Guyana." ExxonMobil. July 28, 2021. https://corporate.exxonmobil.com/News/Newsroom/News-releases/2021/0728_ExxonMobil-announces-oil-discovery-at-Whiptail-offshore-Guyana

[7]"Stabroek Block Resources Jump to 9 Billion Barrels of Oil Equivalent – Hess." OilNOW. October 28, 2020. https://oilnow.gy/featured/stabroek-block-resources-jump-to-9-billion-barrels-of-oil-equivalent-hess

[8]"ExxonMobil to Proceed With Payara Development Offshore Guyana." ExxonMobil. September 30, 2020. https://corporate.exxonmobil.com/News/Newsroom/News-releases/2020/0930_ExxonMobil-to-proceed-with-Payara-development-offshore-Guyana

[9]"ExxonMobil Announces Discovery at Uaru-2 Offshore Guyana." ExxonMobil. April 27, 2021. https://corporate.exxonmobil.com/News/Newsroom/News-releases/2021/0427_ExxonMobil-announces-discovery-at-Uaru-2-offshore-Guyana

[10]Prableen Bajpai. "The Five Fastest Growing Economies in the World." Nasdaq. October 16, 2020. https://www.nasdaq.com/articles/the-five-fastest-growing-economies-in-the-world-2020-10-16

[11]Alexis Rodney. "Guyana Trying To Avoid 'Dutch Disease'." Department of Public Information, Guyana. January 20, 2021. https://dpi.gov.gy/guyana-trying-to-avoid-dutch-disease

[12]Lisa Hamilton. "Bynoe Talks up Benefits of Local Content." Guyana Chronicle. December 20, 2019. https://guyanachronicle.com/2019/12/20/bynoe-talks-up-benefits-of-local-content

[13]Anara Khan. "Local Content a Priority for New PPP/C Gov't." Department of Public Information, Guyana. August 16, 2020. https://dpi.gov.gy/local-content-a-priority-for-new-ppp-c-govt

[14]"Guyanese Taxpayers Will Have To Foot the Bill for Any Inefficiencies Triggered by Local Content Rules – Floyd Haynes." OilNOW. March 20, 2021. https://oilnow.gy/featured/guyanese-taxpayers-will-have-to-foot-the-bill-for-any-inefficiencies-triggered-by-local-content-rules-floyd-haynes

[15]"Operators Must Tell Local Businesses About Upcoming Opportunities, President Says." OilNOW. February 16, 2021. https://oilnow.gy/featured/operators-must-tell-local-businesses-about-upcoming-opportunities-president-says

Chapter 18: Diversifying Energy and Diversifying Economies

[1]Vladimir Lenin. "Our Foreign and Domestic Position and Party Tasks." Speech Delivered to the Moscow Gubernia Conference Of The R.C.P.(B.). November 21, 1920. https://www.marxists.org/archive/lenin/works/1920/nov/21.htm

[2]Jerome Blum. "Russian Agriculture in the Last 150 Years of Serfdom." Agricultural History. January 1960. http://blogs.bu.edu/guidedhistory/files/2012/09/Russian-Agriculture-in-the-Last-150-Years-of-Serfdom.pdf

[3]Anton Cheremukhin, Mikhail Golosov, Sergei Guriev, and Aleh Tsyvinski. "The Industrialization and Economic Development of Russia Through the Lens of a Neoclassical Growth Model." Princeton. November 2015. https://scholar.princeton.edu/sites/default/files/golosov/files/cggt_revision.pdf

[4]Vladimir Lenin. "Our Foreign and Domestic Position and Party Tasks." Speech Delivered to the Moscow Gubernia Conference Of The R.C.P.(B.). November 21, 1920. https://www.marxists.org/archive/lenin/works/1920/nov/21.htm

[5]"Oil and Gas Challenges and Opportunities in Africa." GEO ExPro. 2019. https://www.geoexpro.com/articles/2019/09/oil-and-gas-challenges-and-opportunities-in-africa

[6]Manfred Hafner, Simone Tagliapietra, and Lucia de Strasser. "Prospects for Renewable Energy in Africa." Energy in Africa. 2018. https://doi.org/10.1007/978-3-319-92219-5_3

[7]Sergio Chapa. "Toyota, Kia Plants in Mexico To Fuel Demand for Eagle Ford Natural Gas." San Antonio Business Journal. April 6, 2015. https://www.bizjournals.com/sanantonio/blog/eagle-ford-shale-insight/2015/04/toyota-kia-plants-in-mexico-to-fuel-demand-for.html

[8]"Senegal's Karpowership Secures US$60m From Mauritius Commercial Bank." CEO Business Africa. April 23, 2021. https://ceobusinessafrica.com/senegals-karpowership-secures-us60m-from-mauritius-commercial-bank

[9]J. Peter Pham. "Africa's Real Strategic Import for the Green Economy." Atlantic Council. March 29, 2021. https://www.atlanticcouncil.org/blogs/africasource/africas-real-strategic-import-for-the-green-economy

[10]Catherine Rollet. "Are Rare Earths Used in Solar Panels?" PV Magazine. November 28, 2019. https://www.pv-magazine.com/2019/11/28/are-rare-earths-used-in-solar-panels

[11]Abdi Latif Dahir. "Africa's Largest Wind Power Project Is Now Open in Kenya." Quartz Africa. July 22, 2019. https://qz.com/africa/1671484/kenya-opens-africas-largest-wind-power-project-in-turkana

[12]"Virginia Gas Project." Renergen. n.d. https://www.renergen.co.za/renergen-projects/virginia-gas-project

[13]"Virginia Gets Own Liquefied Gas Plant." Bloemfontein Courant. November 27, 2019. https://www.bloemfonteincourant.co.za/virginia-gets-own-liquefied-gas-plant

[14]"Renergen Signs LNG Supply Agreement With Logico." Renergen. n.d. https://www.renergen.co.za/renergen-asxrlt-signs-lng-supply-agreement-with-logico

[15]Bruno Merven, Faaiqa Hartley, and Fadiel Ahjum. "Road Freight and Energy in South Africa." Southern Africa –Towards Inclusive Economic Development (SA-TIED). 2018. https://sa-tied.wider.unu.edu/sites/default/files/pdf/SATIED_WP60_Merven_Hartley_Ahjum_April_2019.pdf

[16]"The Future of Road Freight Transport." FDL Transport. n.d. https://fdktransport.co.za/future-road-freight-transport

[17]Liz Kneale. "Freight Transport by Road in South Africa 2018. Who Owns Whom. January 25, 2018. https://www.whoownswhom.co.za/store/info/4565?segment=

[18]Bruno Merven, Faaiqa Hartley, and Fadiel Ahjum. "Road Freight and Energy in South Africa." Southern Africa –Towards Inclusive Economic Development (SA-TIED). 2018. https://sa-tied.wider.unu.edu/sites/default/files/pdf/SATIED_WP60_Merven_Hartley_Ahjum_April_2019.pdf

[19]"Road Freight." ExportHelp. 2022. https://exporthelp.co.za/modules/16_logistics/road_freight.html

[20]"Almost Half of South Africa's Oil Refining Seen Shut Until 2022." Bloomberg. January 11, 2021. https://businesstech.co.za/news/energy/459932/almost-half-of-south-africas-oil-refining-seen-shut-until-2022

[21]Simone Liedtke. "Home-Grown Gas Innovation Set To Deliver Zero-Emission Refrigerated Transport." Creamer Media. December 4, 2020. https://www.engineeringnews.co.za/article/foot-on-the-gas-2020-12-04/rep_id:4136

[22]"How Many Jobs Can Dangote Refinery Actually Create?" Arbeiterz. November 16, 2020. https://arbiterz.com/how-many-jobs-can-dangote-refinery-actually-create

[23]Dario Giuliani and Sam Ajadi. "618 Active Tech Hubs: The Backbone of Africa's Tech Ecosystem." GSMA. July 10, 2019. https://www.gsma.com/mobilefordevelopment/blog/618-active-tech-hubs-the-backbone-of-africas-tech-ecosystem

[24]"Is IronRidge's Ewoyaa in Ghana a Leading Lithium Asset?" IronRidge Resources. January 22, 2021. https://www.miningreview.com/battery-metals/is-ironridge-resources-ewoyaa-in-ghana-a-leading-lithium-asset

[25]"RCL Foods: Customer Story Key Figures." Schneider Electric. 2022. https://www.se.com/us/en/work/campaign/life-is-on/case-study/rcl-foods.jsp

[26]"Sahara Wind Energy Potential: Powering Africa & Europe." Sahara Wind Inc. 2022. https://saharawind.com/sahara-wind-energy-potential-powering-africa-and-europe

[27]"Senegal: Cretaceous Play Extension Opportunity." Kosmos. n.d. https://www.kosmosenergy.com/senegal

[28]"Senegal Already Produces Gas: Investors Should Look Onshore." African Energy Chamber. January 15, 2021. https://www.accesswire.com/624461/Senegal-Already-Produces-Gas-Investors-Should-Look-Onshore

[29]"USTDA Supports Senegal's First Major Gas Pipeline." US Trade and Development Agency. August 28, 2020. https://ustda.gov/ustda-supports-senegals-first-major-gas-pipeline

[30]Sanja Pekic. "MCB Bankrolls Senegal's Use of Floating LNG Power Plants. Offshore Energy. April 23, 2021. https://www.offshore-energy.biz/mcb-bankrolls-senegals-use-of-floating-lng-power-plants

[31]"The World Bank in Senegal." The World Bank. n.d. https://www.worldbank.org/en/country/senegal/overview

[32]"GE Secures Equipment Contract to Power the Biggest Power Plant in Senegal." January 28, 2021. https://www.ge.com/news/press-releases/ge-secures-equipment-contract-to-power-the-biggest-power-plant-in-senegal

[33]Aisha Salaudeen. "Senegal Launches Wind Power Plant as Part of its Commitment to Renewable Energy." CNN. February 25, 2020. https://www.cnn.com/2020/02/25/africa/west-africas-largest-wind-plant/index.html

[34]"Senegal: Power Africa Fact Sheet." USAID. April 5, 2022. https://www.usaid.gov/powerafrica/senegal

Chapter 19: AI, Analytics, and Innovation: How Africa Can Make the Leap to New Technologies

[1]"Leapfrogging." Wikipedia. June 1, 2022. https://en.wikipedia.org/wiki/Leapfrogging

[2]"Mobile Technology in Africa." Wikipedia. August 30, 2021. https://en.wikipedia.org/wiki/Mobile_technology_in_Africa

[3]Eliza Strickland. "With "Leapfrog" Technologies, Africa Aims to Skip the Present and Go Straight to the Future." IEEE Spectrum. April 30, 2019. https://spectrum.ieee.org/robotics/drones/with-leapfrog-technologies-africa-aims-to-skip-the-present-and-go-straight-to-the-future

[4]"M-Pesa: When "Mobile Money" Revolutionizes Banking in Africa." Harvard Business School. November 18, 2016. https://digital.hbs.edu/platform-rctom/submission/m-pesa-when-mobile-money-revolutionizes-banking-in-africa

[5]"M-Pesa." Wikipedia. May 19, 2022. https://en.wikipedia.org/wiki/M-Pesa

[6]Isabelle Gerretsen. "Trucking App Kobo360 Wants To Halve Delivery Times Across Africa. CNN Business. April 16, 2020. https://www.cnn.com/2020/04/16/tech/kobo-360-trucks-spc-intl/index.html

[7]"Cell Phones Allow Countries to 'Leapfrog' Technology." VOA. November 1, 2009. https://www.voanews.com/a/a-13-2008-05-19-voa22/401756.html

[8]Isabelle Gerretsen. "Trucking App Kobo360 Wants To Halve Delivery Times Across Africa. CNN Business. April 16, 2020. https://www.cnn.com/2020/04/16/tech/kobo-360-trucks-spc-intl/index.html

[9]Amit Patil. "Artificial Intelligence and Big Data in the Petrochemical Sector – Probing and Imaging Technology: Then and Now." SINTEF. April 27, 2018. https://blog.sintef.com/industry-en/z-big-data-petrochemical-sector-probing-imaging-technology-then-now

[10]"4 Benefits of AI in the Operation of Power Plants." Uniper. July 23, 2020. https://www.uniper.energy/news/4-benefits-of-ai-in-the-operation-of-power-plants

[11]Nicholas Nhede. "Australia's First AI-Enabled Wind Farm." Smart Energy International. September 19, 2019. https://www.smart-energy.com/renewable-energy/australias-first-ai-enabled-wind-farm

[12]"Machine Learning & Solar Technology." IGS Energy. n.d. https://www.igs.com/energy-resource-center/energy-101/machine-learning-solar-technology

[13]"Next-Gen Fuel Retail Solutions for Convenience Stores." ORPAK. 2022. https://www.orpak.com/solutions/forevision-business-insight-optimization

[14]"Repsol To Optimize Refinery Management Through Big Data and Artificial Intelligence With Google Cloud." Repsol. June 4, 2018. https://www.repsol.com/imagenes/global/en/repsol_google_cloud_ing_tcm14-132376.pdf

[15]"Predictive Pipeline Data Analysis With Integrated Applications, Planning & Analysis Software." Mistras. n.d. https://www.mistrasgroup.com/who-we-help/industries/oil-and-gas/midstream/pipelines/pipeline-predictive-analytic-solutions

[16]"SpeedWise Galaxy." Quantum Reservoir Impact, LLC. n.d. https://www.qrigroup.com/SpeedWise

[17]"Automate Visual Forensics with BitBox." Teradata. 2020. https://assets.teradata.com/resourceCenter/downloads/Datasheets/Automate-Visual-Forensics-with-BitBox-MD005455.pdf

[18]Naveen Joshi. "Expenditures in an IoT Initiative." Allerin. May 19, 2017. https://www.allerin.com/blog/expenditures-in-an-iot-initiative

[19]Mark Stevens. "6 Factors That Determine IoT Price for Manufacturers." WIPFLI. September 14, 2021. https://www.wipfli.com/insights/blogs/manufacturing-tomorrow-blog/180515---cost-to-implement-iot

[20]Jennifer Ann Patterson. "Poor Technological Capability Undermining Africa's Growth Potential." African Development Bank Group. November 1, 2014. https://www.afdb.org/en/news-and-events/poor-technological-capability-undermining-africas-growth-potential-13684

[21]"Economic Development in Africa: Report 2019." United Nations Conference on Trade and Development. 2019. https://unctad.org/system/files/official-document/aldcafrica2019_en.pdf

[22]Mfonobong Nsehe. "Ten Minutes With Jobberman.com Cofounder Ayodeji Adewunmi. Forbes. September 5, 2011. https://www.forbes.com/sites/mfonobongnsehe/2011/09/05/ten-minutes-with-jobberman-com-cofounder-ayodeji-adewunmi/?sh=6b9857305cd2

[23]Madanmohan Rao. "Startup Guide Nairobi: How 'Silicon Savannah' Is a Hub of Creativity, Tech and Impact." YourStory. January 24, 2021. https://yourstory.com/2021/01/startup-guide-nairobi/amp

[24]Economic Development in Africa: Report 2019." United Nations Conference on Trade and Development. 2019. https://unctad.org/system/files/official-document/aldcafrica2019_en.pdf

[25]Elijah Dickens Mushemeza and John Okiira. "Local Content Frameworks in the African Oil and Gas Sector: Lessons from Angola and Chad." Evidence and Lessons from Latin America. April 2016. http://ella.practicalaction.org/wp-content/uploads/2016/07/REP-ACODE-Local-Content Frameworks-in-the-African-Oil-and-Gas.pdf

[26]Klaus Schwab. "The Fourth Industrial Revolution: What it Means, How To Respond." World Economic Forum. January 14, 2016. https://www.weforum.org/agenda/2016/01/the-fourth-industrial-revolution-what-it-means-and-how-to-respond

[27]Jeff Dean and Moustapha Cisse. "Google AI in Ghana." Google in Africa. June 13, 2018. https://blog.google/around-the-globe/google-africa/google-ai-ghana

[28]Kent Mensah. "Google Takes on 'Africa's Challenges' With First AI Centre in Ghana. Phys.org. April 13, 2019. https://phys.org/news/2019-04-google-africa-ai-centre-ghana.html

Chapter 20: Cobalt Blues: Can We Keep the EV Revolution from Exploiting Africa?

[1]Patrick Hillberg and Sawyer Hall. "How Carmakers' Switch to Electric Vehicles Will Strain Supply of Battery Minerals." World Economic Forum. June 25, 2021. https://www.weforum.org/agenda/2021/06/carmakers-switch-to-electric-vehicles-strain-supply-of-battery-minerals

[2]Simon Wilson. "How China Is Cornering the Market for Electric-Car Batteries." MoneyWeek. January 8, 2022. https://moneyweek.com/investments/commodities/industrial-metals/604306/china-cobalt-electric-car-batteries

[3]Patrick Hillberg and Sawyer Hall. "How Carmakers' Switch to Electric Vehicles Will Strain Supply of Battery Minerals." World Economic Forum. June 25, 2021. https://www.weforum.org/agenda/2021/06/carmakers-switch-to-electric-vehicles-strain-supply-of-battery-minerals

[4]"Electric Vehicle Sales Headed for Five and a Half Million In 2021 as Automakers Target 40 Million Per Year by 2030." BloombergNEF. November 10, 2021. https://about.bnef.com/blog/electric-vehicle-sales-headed-for-five-and-a-half-million-in-2021-as-automakers-target-40-million-per-year-by-2030

[5]"Battery Innovation Must Drive the 50-Fold Increase in Storage Capacity Needed by 2040." International Energy Agency. November 6, 2020. https://energypost.eu/battery-innovation-must-drive-the-50-fold-increase-in-storage-capacity-needed-by-2040

[6]"Battery Innovation Must Drive the 50-Fold Increase in Storage Capacity Needed by 2040." International Energy Agency. November 6, 2020. https://energypost.eu/battery-innovation-must-drive-the-50-fold-increase-in-storage-capacity-needed-by-2040

[7]Patrick Hillberg and Sawyer Hall. "How Carmakers' Switch to Electric Vehicles Will Strain Supply of Battery Minerals." World Economic Forum. June 25, 2021. https://www.weforum.org/agenda/2021/06/carmakers-switch-to-electric-vehicles-strain-supply-of-battery-minerals

[8]M. Garside. "Cobalt Reserves Worldwide as of 2021, by Country." Statista. Mar 11, 2022. https://www.statista.com/statistics/264930/global-cobalt-reserves/#:~:text=The%20Democratic%20Republic%20of%20the,world's%20reserves%20of%20the%20metal

[9]Rafiq Raji. "Electric Vehicles: Africa's Battery Minerals and GVC Opportunities." Nanyang Technological University. August 13, 2021. https://www.ntu.edu.sg/cas/news-events/news/details/electric-vehicles-africa-s-battery-minerals-and-gvc-opportunities

[10]Simon Wilson. "How China Is Cornering the Market for Electric-Car Batteries." MoneyWeek. January 8, 2022. https://moneyweek.com/investments/commodities/industrial-metals/604306/china-cobalt-electric-car-batteries

[11]John Xie. "How China Dominates Global Battery Supply Chain. VOA. September 1, 2020. https://www.voanews.com/a/silicon-valley-technology_how-china-dominates-global-battery-supply-chain/6195257.html

[12]Govind Bhutada. "Mapped: EV Battery Manufacturing Capacity, by Region." Visual Capitalist. February 28, 2022. https://www.visualcapitalist.com/mapped-ev-battery-manufacturing-capacity-by-region

[13]Jacky Wong. "EV Makers' Next Headache: Scarce Battery Chemicals, Made in China." January 21, 2022. https://www.wsj.com/articles/ev-makers-next-headache-scarce-battery-chemicals-made-in-china-11642768194#:~:text=China%20in%20general%20has%20more,the%20world's%20natural%20flake%20graphite

[14]Dionne Searcey, Michael Forsythe, and Eric Lipton. "A Power Struggle Over Cobalt Rattles the Clean Energy Revolution." The New York Times. December 7, 2021. https://www.nytimes.com/2021/11/20/world/china-congo-cobalt.html

[15]Halim Nazar. "Chinese Mining in the DRC: From Sicomines to Global Cobalt Monopoly." The Institute of Chinese Studies. August 27, 2021. https://icsin.org/blogs/2021/08/27/chinese-mining-in-the-drc-from-sicomines-to-global-cobalt-monopoly/#:~:text=Under%20the%20agreement%2C%20China%20would,infrastructure%20and%20mining%20credits%20combined

[16]Dionne Searcey, Michael Forsythe, and Eric Lipton. "A Power Struggle Over Cobalt Rattles the Clean Energy Revolution." The New York Times. December 7, 2021. https://www.nytimes.com/2021/11/20/world/china-congo-cobalt.html

[17]"Programme Sino-Congolais." Agence Congolaise des Grands Travaux (ACGT). 2022. http://www.acgt.cd/projets/programme-sino-congolais

[18]Pete Pattisson. "'Like Slave and Master': DRC Miners Toil for 30p an Hour To Fuel Electric Cars." The Guardian. November 8, 2021. https://www.theguardian.com/global-development/2021/nov/08/cobalt-drc-miners-toil-for-30p-an-hour-to-fuel-electric-cars

[19]Pete Pattisson. "'Like Slave and Master': DRC Miners Toil for 30p an Hour To Fuel Electric Cars." The Guardian. November 8, 2021. https://www.theguardian.com/global-development/2021/nov/08/cobalt-drc-miners-toil-for-30p-an-hour-to-fuel-electric-cars

[20]Pete Pattisson. "'Like Slave and Master': DRC Miners Toil for 30p an Hour To Fuel Electric Cars." The Guardian. November 8, 2021. https://www.theguardian.com/global-development/2021/nov/08/cobalt-drc-miners-toil-for-30p-an-hour-to-fuel-electric-cars

[21]"'This Is What We Die For': Human Rights Abuses in the Democratic Republic of the Congo Power the Global Trade in Cobalt." Amnesty International. 2016. https://www.amnesty.org/en/wp-content/uploads/2021/05/AFR6231832016ENGLISH.pdf

[22]Dionne Searcey, Michael Forsythe, and Eric Lipton. "A Power Struggle Over Cobalt Rattles the Clean Energy Revolution." The New York Times. December 7, 2021. https://www.nytimes.com/2021/11/20/world/china-congo-cobalt.html

[23]"'This Is What We Die For': Human Rights Abuses in the Democratic Republic of the Congo Power the Global Trade in Cobalt." Amnesty International. 2016. https://www.amnesty.org/en/wp-content/uploads/2021/05/AFR6231832016ENGLISH.pdf

[24]"'This Is What We Die For': Human Rights Abuses in the Democratic Republic of the Congo Power the Global Trade in Cobalt." Amnesty International. 2016. https://www.amnesty.org/en/wp-content/uploads/2021/05/AFR6231832016ENGLISH.pdf

[25]Remeredzai Joseph Kuhudzai. "New South African Lithium-Ion Cell Mega-Factory, Plans For 32 GWh/Year By 2028." Clean Technica. February 9, 2020. https://cleantechnica.com/2020/02/09/new-south-african-lithium-ion-cell-mega-factory-plans-for-32-gwh-year-by-2028

[26]"Infographic: Australia Mining by the Numbers." S&P Global. February 12, 2022. https://www.spglobal.com/marketintelligence/en/news-insights/blog/infographic-australia-mining-by-the-numbers

[27]Elisha Newell. "Magnis Energy Technologies Advances Nachu Graphite Project into 2022." Proactive. February 10, 2022. https://www.proactiveinvestors.com/companies/news/973841/magnis-energy-technologies-advances-nachu-graphite-project-into-2022-973841.html

[28]"Battery Minerals, Urbix Partner for Graphite Processing JV in Mozambique." NS Energy. October 10, 2019. https://www.nsenergybusiness.com/news/battery-minerals-graphite-processing-jv

[29]"Cobalt-Rich Congo Tries to Push into Battery Manufacturing." Reuters. November 25, 2021. https://www.reuters.com/markets/commodities/cobalt-rich-congo-tries-push-into-battery-manufacturing-2021-11-24

[30]"The Role of Critical Minerals in Clean Energy Transitions." International Energy Agency. May 2021. https://www.iea.org/reports/the-role-of-critical-minerals-in-clean-energy-transitions/executive-summary

[31]Tariye Gbadegesin. "Africa Must Assume its Place in the Global Battery Race." The Africa Report. January 7, 2020. https://www.theafricareport.com/21865/africa-must-assume-its-place-in-the-global-battery-race

[32]Anmar Frangoul. "VW and Goldman Sachs-Backed Northvolt Plans German Gigafactory. CNBC. March 15, 2022. https://www.cnbc.com/2022/03/15/vw-and-goldman-sachs-backed-northvolt-plans-german-gigafactory.html

[33]Scooter Doll. "How the US Plans To Capture the EV Battery Market." Electrek. February 5, 2021. https://electrek.co/2021/02/05/how-the-us-plans-to-capture-the-ev-battery-market

[34]"Women's Economic Empowerment in Sub-Saharan Africa." BSR. March 2017. https://www.bsr.org/reports/BSR_Womens_Empowerment_Africa_Main_Report.pdf

[35]Local Content Policies in Minerals-Exporting Countries, Case Studies. Organisation for Economic Cooperation and Development (OECD). June 2, 2017. https://www.oecd.org/officialdocuments/publicdisplaydocumentpdf/?cote=TAD/TC/WP(2016)3/PART2/FINAL&docLanguage=En

Chapter 21: Could Namibia Serve as a Template for Energy Transition Success in Africa?

[1]"Kudu Gas Field." Wikipedia. September 14, 2020. https://en.wikipedia.org/wiki/Kudu_gas_field

[2]Ron Bousso and Wendell Roelf. "Shell Hits Oil and Gas in Namibian Offshore Well." Reuters. January 25, 2022. https://www.reuters.com/business/energy/exclusive-shell-hits-oil-gas-namibian-offshore-well-2022-01-25

[3]Iain Esau. "Shell Poised to Spud Appraisal Well on Potentially Huge Graff Oil Discovery in Namibia." Upstream. February 16, 2022. https://www.upstreamonline.com/exclusive/shell-poised-to-spud-appraisal-well-on-potentially-huge-graff-oil-discovery-in-namibia/2-1-1168213

[4]Iain Esau. "At Least 3 Billion Barrels: TotalEnergies' Venus Is Sub-Saharan Africa's Biggest Ever Oil Discovery." Upstream. February 28, 2022. https://www.upstreamonline.com/exclusive/at-least-3-billion-barrels-totalenergies-venus-is-sub-saharan-africas-biggest-ever-oil-discovery/2-1-1174983

[5]"Oil Price Charts." OilPrice.com. n.d. https://oilprice.com/oil-price-charts

[6]Brian Kenety. "Namibia to Fast-Track Development of Offshore Oil Discovery at Graff-1." Newsbase. February 9, 2022. https://newsbase.com/story/namibia-to-fast-track-development-of-offshore-oil-discovery-at-graff-1-234477

[7]Iain Esau. "At Least Three FPSOs on Cards as Namibia Urges Fast-Track Development of Venus and Graff." March 14, 2022. https://www.upstreamonline.com/field-development/at-least-three-fpsos-on-cards-as-namibia-urges-fast-track-development-of-venus-and-graff/2-1-1181848

[8]"Our Ambition With Fast4Ward®." Offshore. n.d. https://www.sbmoffshore.com/creating-value/fast4wardr

[9]"Gas, Water Depths Could Complicate Namibian Field Developments." Offshore. April 8, 2022. https://www.offshore-mag.com/regional-reports/africa/article/14270822/gas-water-depths-could-complicate-namibian-field-developments

[10]Fabio Scala. "Namibia kickstarts oil boom with sizable 2022 discoveries," Further Africa, April 22, 2022, https://furtherafrica.com/2022/04/22/namibia-kickstarts-oil-boom-with-sizeable-2022-discoveries

[11]Hage G. Geingob. "Namibia Is Poised to Become the Renewable Energy Hub of Africa." World Economic Forum. October 3, 2021. https://www.weforum.org/agenda/2021/10/namibia-is-positioned-to-become-the-renewable-energy-hub-of-africa

[12]"Namibia: Power Africa Fact Sheet." USAID. November 18, 2021. https://www.usaid.gov/powerafrica/namibia

[13]"Namibia Announces $9.4 Billion Green Hydrogen Project." FuleCellsWorks. November 5, 2021. https://fuelcellsworks.com/news/namibia-announces-9-4-billion-green-hydrogen-project

[14]"GDP (Current US$) – Namibia." The World Bank. n.d. https://data.worldbank.org/indicator/NY.GDP.MKTP.CD?locations=NA

[15]Anmar Frangoul. "Shell Says One of the Largest Hydrogen Electrolyzers in the World Is Now Up and Running in China." CNBC. January 28, 2022. https://www.cnbc.com/2022/01/28/green-hydrogen-one-of-planets-largest-electrolyzers-up-and-running.html

[16]Tasneem Bulbulia. "South Africa, Namibia Among New African Green Hydrogen Alliance's Founding Members. Creamer Media. May 18, 2022. https://www.engineeringnews.co.za/article/south-africa-namibia-among-new-african-green-hydrogen-alliances-founding-members-2022-05-18/rep_id:4136

[17]Tembile Sgqolana. "Shell's Oil Discovery in Namibia Won't Benefit Namibians, Say Activists." Daily Maverick. February 23, 2022. www.dailymaverick.co.za/article/2022-02-23-shells-oil-discovery-in-namibia-wont-benefit-namibians-say-activists

[18]Hage G. Geingob. "Namibia Is Poised to Become the Renewable Energy Hub of Africa." World Economic Forum. October 3, 2021. https://www.weforum.org/agenda/2021/10/namibia-is-positioned-to-become-the-renewable-energy-hub-of-africa

[19]"Household Air Pollution and Health." World Health Organization. September 22, 2021. https://www.who.int/news-room/fact-sheets/detail/household-air-pollution-and-health

[20]"Converting Natural Gas Pipelines to Hydrogen." KNF. January 12, 2022. https://knf.com/en/se/stories-events/blog/knowledge/article/hydrogen-natural-gas-pipelines

[21]"Hydrogen Pipelines." US Department of Enegry Office of Energy Efficiency & Renewable Energy. n.d. https://www.energy.gov/eere/fuelcells/hydrogen-pipelines

[22]Jeff St. John. "Green Hydrogen in Natural Gas Pipelines: Decarbonization Solution or Pipe Dream?" Greentech Media. November 30, 2020. https://www.greentechmedia.com/articles/read/green-hydrogen-in-natural-gas-pipelines-decarbonization-solution-or-pipe-dream